Proceedings of the 3rd International Colloquium on Sports Science, Exercise, Engineering and Technology

Norasrudin Sulaiman · Shariman Ismadi Ismail
Rahmat Adnan
Editors

Proceedings of the 3rd International Colloquium on Sports Science, Exercise, Engineering and Technology

ICoSSEET 2016, 20–22 November 2016,
Kota Kinabalu, Malaysia

Springer

Editors
Norasrudin Sulaiman
Faculty of Sports Science and Recreation
Universiti Teknologi MARA
Shah Alam, Selangor, Malaysia

Rahmat Adnan
Faculty of Sports Science and Recreation
Universiti Teknologi MARA
Shah Alam, Selangor, Malaysia

Shariman Ismadi Ismail
Faculty of Sports Science and Recreation
Universiti Teknologi MARA
Shah Alam, Selangor, Malaysia

ISBN 978-981-10-6771-6 ISBN 978-981-10-6772-3 (eBook)
https://doi.org/10.1007/978-981-10-6772-3

Library of Congress Control Number: 2018950777

This Springer imprint is published by the registered company Springer Nature Singapore Pte Ltd.
The registered company address is: 152 Beach Road, #21-01/04 Gateway East, Singapore 189721, Singapore

Contents

Part I
Sports and Exercise Science

Moderating Effects of Gender and Types of Sport on the Relationships Between Coaching Leadership, Athletes' Psychological Needs, and Motivation Among University Athletes in Malaysia

Kang Mea Kee and Anis Syaimaa Husna Ismail

Abstract The purpose of this study was to examine the moderation effects of gender on the relationships between the athletes' perception of coach's leadership (CL) and athletes' basic psychological needs (BPN) and the moderation effect of types of sports in relation to their sport motivation (AM) level among university athletes. A total of 350 university athletes, consisting of 199 males and 151 females from 9 contact sports and 14 noncontact sports participated in this study. The measurement model was first tested for internal reliability, construct reliability, convergent validity, and discriminant validity for all constructs. The results of structural model showed that athletes' perception of coaches' leadership was significantly related with athletes' basic psychological needs, and subsequently, athletes' basic psychological needs were significantly related with athletes' motivation. PLS-MGA results indicated that there was a significant moderation effect of gender on relationship between athletes' perception of coach leadership and athletes' psychological needs. Similarly, the types of sports significantly moderated the relationship between athletes' psychological needs and athletes' motivation. As a conclusion, coaches' leadership style has an influence on the psychological needs satisfaction of the athletes, and this subsequently affects athletes' motivation. However, these relationships were being moderated by gender and types of sports, respectively. Hence, coaches must take into consideration the gender factor and the types of sports when training the athletes.

Keywords Coaches' leadership · Athletes' basic psychological needs · Athletes' motivation · Gender · Types of sports

---

K. M. Kee (✉) · A. S. H. Ismail
Faculty of Sport Science and Recreation, Universiti Teknologi MARA, Shah Alam, Malaysia
e-mail: kee@salam.uitm.edu.my

N. Sulaiman et al. (eds.), *Proceedings of the 3rd International Colloquium on Sports Science, Exercise, Engineering and Technology*,
https://doi.org/10.1007/978-981-10-6772-3_1

1 Introduction

Athletes have the desire to be the best and successful in their sports. In order to be successful, they often have to go through great sacrifices such as time with family, friends, and even personal time. Yet these personal sacrifices are not the only factors that determine athletes' success. Some studies said that social environment such as coach leadership has a significant impact on athletes' success [1, 2]. Effective coaching leadership emphasizes on creating a motivational environment that encourages athletes or teams to perform toward their full potential [3]. Coaching leadership has been characterized into five dimensions, namely, training and instruction, democratic behavior, autocratic behavior, social support, and positive feedback [4].

The success of athletes is also influenced by their basic psychological needs. Human psychological needs is a behavior deriving from the individuals' true self which results in higher levels of behavioral self-determination when being satisfied [5]. Deci and Vansteenkiste [6] stated that athletes have a greater sport motivation when all the three psychological needs (need of competence, autonomy, and relatedness) are being satisfied. Effective coaches often create opportunities that help to fulfill athletes' psychological needs and lead them to have a greater self-determination toward positive motivational outcomes such as persistent participation. In addition, the relationship between coaching leadership and motivation is said to being mediated by these three innate psychological human needs. Past studies have shown that motivated athletes exhibit positive relationships with the fulfillment of the need of autonomy, competence, and relatedness [7]. Previous studies have also indicated that these psychological needs are unique between genders, and hence, coaching leadership style might vary for males and females in order to satisfy different psychological needs in sport [8]. In addition, the types of sports might also influence athletes' psychological needs. For example, contact sports may require different mind-set (e.g., more aggressive) as the level of aggression is being projected as part of the characteristics of the contact sport itself. Extrinsically motivated athletes will participate in sport to satisfy external reward such as cash prize. However, the most important reason for athletes being successful in sport is when they are intrinsically motivated (participate in sport for enjoyment and interest), in which athletes are said to show greater endurance, more effort during practices and game, less boredom and showed a higher level of excitement [9].

Most of the studies on coaching leadership have been conducted in advanced or developed countries such as Northern America and Europe. However, it is not clear whether coaches in Malaysia which have very different cultural values will exhibit similar patterns of leadership in sports. In addition, it appears that coaching leadership in sports in Malaysia has not been fully explored. No study has yet been sighted locally that examine the relationship of coaching leadership and psychological needs on athletes' motivation in sport as proposed by Hollembeak and Amorose [10]. Hence, in response to the current issues, the purpose of this study was to examine the relationship between the athletes' perception of coaches' leadership (CL) and athletes' psychological needs (BPN) on the athletes' motivation (AM). Specifically,

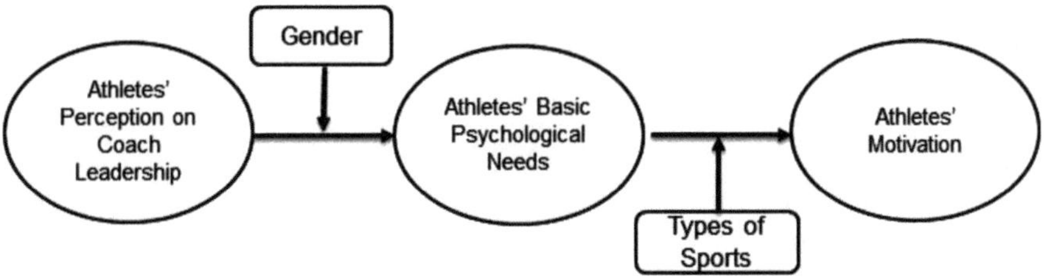

Fig. 1 Conceptual framework of the study

this study also examined if gender and types of sport have the said moderating role in current model involving the three constructs, namely, relationship of coaching leadership and psychological needs and the relationship between athletes' basic psychological needs on athletes' motivation in sport among the Malaysian university athletes.

2 Method

This study employed the correlational research design that utilized the questionnaires technique to collect data. A total of 350 participants comprised of males ($n = 199$) and females ($n = 151$) university athletes, age ranged between 19 and 30 years old, from 9 contact sports, and 14 noncontact sports participated in this study. This study employed the athlete's perception version of Leadership Scale for Sports (LSS, 40 items) questionnaire by Chelladurai and Saleh [4] and participants answered using a 5-point Likert scale. As for Basic Psychological Needs Satisfaction in Sport questionnaire (BPNSS, 20 items) by Ng et al. [11], and Revised Sport Motivation Scale (SMS-II, 18 items) questionnaire by Pelletier et al. [12], the participants answered using a 7-point Likert scale. Descriptive data analysis and Partial Least Square Structural Equation Modeling (PLS-SEM) were used to analyze the data.

Figure 1 shows the conceptual framework of the study involving the three constructs, namely, athletes' perception of coach leadership, athletes' basic psychological needs, and athletes' motivation. The moderation effect of gender and types of sports were examined in the model.

3 Results and Discussion

As shown in Table 1, the highest mean for the overall athletes' perception of coach's leadership was positive feedback ($M = 4.44$, SD $= 0.44$), followed by training and instruction ($M = 4.15$, SD $= 0.46$), social support ($M = 4.05$, SD $= 0.33$), and democratic behavior ($M = 3.84$, SD $= 0.66$). The mean for training and instruction and

Table 1 Means and standard deviations for leadership in sport, athlete basic psychological needs, and athlete motivation of university athletes for gender and types of sports

Variables	Male (*n* = 199)		Female (*n* = 151)		Contact sports (*n* = 180)		Noncontact sports (*n* = *170*)		Overall (*N* = *350*)			
	M	SD	*M*	SD	*M*	SD	*M*	SD	Min	Max	*M*	SD
CL_TI	4.12	0.47	4.19	0.45	4.16	0.43	4.14	0.50	3.00	5.00	4.15	0.46
CL_PF	4.42	0.46	4.46	0.42	4.46	0.42	4.42	0.46	3.20	5.00	4.44	0.44
CL_DB	3.80	0.66	3.90	0.65	3.87	0.55	3.80	0.75	2.00	5.00	3.84	0.66
CL_SS	4.03	0.33	4.08	0.31	4.07	0.30	4.03	0.35	3.25	4.38	4.05	0.33
TOTAL CL	4.06	0.41	4.13	0.39	4.11	0.35	4.07	0.45	2.83	4.63	4.09	0.40
BPN_COM	5.95	0.55	5.91	0.46	5.95	0.46	5.91	0.56	4.80	7.00	5.93	0.51
BPN_AUTO	4.62	0.73	4.73	0.69	4.71	0.64	4.62	0.79	3.10	6.40	4.67	0.71
BPN_RLTD	6.43	0.54	6.39	0.51	6.45	0.50	6.37	0.55	5.00	7.00	6.41	0.53
TOTAL BPN	5.40	0.56	5.44	0.53	5.46	0.48	5.38	0.61	4.00	6.70	5.42	0.55
AM_INT	6.09	0.77	6.06	0.72	6.09	0.78	6.07	0.72	4.67	7.00	6.08	0.75
AM_INTG	5.66	0.78	5.45	0.69	5.61	0.76	5.53	0.74	4.00	7.00	5.57	0.75
AM_IDTF	5.89	0.73	5.78	0.70	5.86	0.74	5.83	0.70	4.33	7.00	5.84	0.72
AM_INTJ	5.69	0.94	5.60	0.88	5.72	0.84	5.57	0.99	3.67	7.00	5.65	0.92
AM_EXT	5.32	1.04	5.28	0.97	5.33	0.97	5.28	1.04	3.67	7.00	5.31	1.01
TOTAL AM	5.73	0.74	5.64	0.68	5.72	0.72	5.65	0.70	4.20	7.00	5.69	0.71

Note CL_TI Training and Instruction, *CL_DB* Democratic Behavior, *CL_AB* Autocratic Behavior, *CL_SS* Social Support, *CL_PF* Positive Feedback, *BPN_COM* Need of Competence, *BPN_AUTO* Need of Autonomy, *BPN_RLTD* Need of Relatedness, *AM_INT* Intrinsic Motivation, *AM_INTG* Integrated Motivation, *AM_IDTF* Identified Motivation, *AM_INTJ* Introjected Motivation, *AM_EXT* Extrinsic Motivation, and *AM_AMOT* A-motivation

positive feedback were slightly higher for males compared to females. As for democratic behavior and social support, the mean was slightly higher for females compared to males. The mean for all variables under athletes' perception of coach's leadership was a little higher for athletes in contact sports compared to athletes in noncontact sports.

In terms of athlete's psychological needs subscales, the overall mean for needs of relatedness ($M = 6.41$, $SD = 0.53$) was the highest, followed by the needs of competence ($M = 5.93$, $SD = 0.51$) and the needs of autonomy ($M = 4.67$, $SD = 0.71$). The mean for the needs of competence and needs of relatedness were slightly higher in males compared to females, while the mean for the needs of autonomy was slightly higher in females compared to males. The mean for all three psychological needs was slightly higher for athletes in contact sports compared to athletes in noncontact sports.

As for motivation, the highest mean for overall athletes' motivation was intrinsic motivation ($M = 6.08$, $SD = 0.75$), followed by identified motivation ($M = 5.58$, $SD = 0.72$), introjected motivation ($M = 5.65$, $SD = 0.92$), integrated motivation

(M =5.57, SD=0.75), and extrinsic motivation (M =5.31, SD=1.01). The mean for all types of motivation was slightly higher in males compared to females. The mean for all types of motivation was slightly higher for athletes in contact sports compared to athletes in noncontact sports.

3.1 Model Testing

Table 2 summarizes the results for Reflective Measurement Model for the all latent constructs. The results show that all the constructs have strong composite reliability indicating a high level of internal consistency of this study.

The average variance extracted (AVE) for all the constructs were also well above the required minimum level of 0.50 [13] indicating a high level of convergent validity. Lastly, Fornell–Larcker criterion provides evidence for the constructs discriminant validity for reflective constructs. Hence, the results of measurement model show that all model evaluation criteria have been met, providing support for the measures' reliability and validity of the model.

Table 2 Result summary for reflective measurement model

Latent variables	Indicators	Loadings (>0.70)	Indicator reliability (>0.5)	CR (0.6–0.7)	AVE (>0.5)	Discriminant validity
Athletes' perception of coach's leadership (CL)	CL_TI	0.787	0.619	0.891	0.673	Yes
	CL_DB	0.892	0.796			
	CL_SS	0.833	0.693			
	CL_PF	0.763	0.582			
Athlete basic's psychological needs (BPN)	BPN_COM	0.848	0.719	0.904	0.758	Yes
	BPN_AUTO	0.836	0.697			
	BPN_RLTD	0.926	0.856			
Athletes' motivation (AM)	AM_INT	0.800	0.666	0.935	0.741	Yes
	AM_INTG	0.828	0.682			
	AM_IDTF	0.926	0.874			
	AM_INTJ	0.833	0.651			
	AM_EXT	0.911	0.828			

Note CL_TI Training and Instruction, *CL_DB* Democratic Behavior, *CL_AB* Autocratic Behavior, *CL_SS* Social Support, *CL_PF* Positive Feedback, *BPN_COM* Need of Competence, *BPN_AUTO* Need of Autonomy, *BPN_RLTD* Need of Relatedness, *AM_INT* Intrinsic Motivation, *AM_INTG* Integrated Motivation, *AM_IDTF* Identified Motivation, *AM_INTJ* Introjected Motivation, *AM_EXT* Extrinsic Motivation, and *AM_AMOT* A-motivation

3.2 *Path Coefficient*

As shown in Table 3, the path coefficient result for (1) athletes' perception of coaches' leadership and athletes' psychological needs ($\beta s = 0.806$, $p < 0.001$); and for (2) athletes' psychological needs and athletes' motivation ($\beta s = 0.857$, $p < 0.001$) were significant. The result indicates that athletes' perception of coaches' leadership has a statistically significant strong positive relationship with athletes' basic psychological needs. This indicates that athletes in this study perceived their coaches' leadership highly. Similarly, the athletes' psychological needs showed a statistically significant strong positive relationship with athletes' motivation. This indicates they are highly motivated in practicing their sports. This finding suggests that when athletes' basic psychological needs are being largely fulfilled, they are also highly motivated in practicing their sports.

3.3 *Multigroup Analysis*

Table 4 shows path coefficient results of multigroup analysis and the results show there was significant moderation effect of gender for the relationship between athletes' perception of coach's leadership and athletes' psychological needs (βs diff= 0.107, $p < 0.013$).

The path coefficient for relationship between coach's leadership style and basic psychological needs is significantly higher for male athletes compared to female athletes. Male athletes might prefer coaches who emphasize more autocratic behavior

Table 3 Results of path coefficients and *t*-statistics

Hypothesis	Pathway	Path coefficients (βs)	*t* values	Sig.	*p* values	Decision
H1	CL → BPN	0.806	33.455	***	0.001	H1 Accepted
H2	BPN → AM	0.857	74.706	***	0.001	H2 Accepted

***Significant at the $p < 0.001$ level

Table 4 Multigroup PLS analysis comparison test results for path coefficient (gender)

Path	Group 1: Male ($n = 199$)		Group 2: Female ($n = 151$)		Group 1 versus Group 2	
	$p1$	se($p(1)$)	$p2$	se($p(2)$)	$\|p(1) - p(2)\|$	p
CL → BPN	0.853	0.024	0.747	0.045	0.107	0.013**

Note $p(1)$ and $p(2)$ are path coefficients of group 1 and group 2; se($p(1)$) and se($p(2)$) are the standard error of $p(1)$ and $p(2)$, respectively

**$p < 0.05$ level

Table 5 Multigroup PLS analysis comparison test results for path coefficient (types of sports)

Path	Group 1: Contact sport (n = 180)		Group 2: Noncontact sport (n = 170)		Group 1 versus Group 2	
	$p1$	se(p(1))	$p2$	se(p(2))	$\lvert p(1) - p(2) \rvert$	p
BPN → AM	0.893	0.012	0.846	0.018	0.047	0.010**

Note $p(1)$ and $p(2)$ are path coefficients of group 1 and group 2; se(p(1)) and se(p(2)) are the standard error of $p(1)$ and $p(2)$, respectively
**$p < 0.05$ level

in order to satisfy their psychological needs for competence, autonomy, and relatedness. This makes sense considering that to be successful and to get the most out of each athlete, coaches need to push athletes out of their comfort zone. This finding is consistent with Beam, Serwatka, and Wilson [14] that indicates that male athletes preferred more autocratic coaching behaviors because successful athletes and teams associated with coaches who are more authoritarian have highest win-loss records.

Table 5 shows PLS-MGA analysis for athletes in contact sports and noncontact sports and the result indicates that types of sports significantly moderate the relationship between athletes' psychological needs and athletes' motivation (βs diff = 0.047, $p < 0.010$).

The path coefficient for relationship between athletes' psychological needs and athletes' motivation was significantly higher in athletes in contact sports compared to athletes in noncontact sports. Athletes in team sports are said to be highly motivated when their psychological need for relatedness is been satisfied [2]. Apparently, most athletes in the contact sports in this study were mostly from team sports. Hence, contact sport athletes tend to be exposed to teamwork dynamic that encourages their feeling of belongingness and related with each teammate. On the other hand, most noncontact sports in this study tend to be individual sports. Since they play as individuals, they tend to be more responsible for their own behaviors such as determine their training schedule and choosing which competitions they want to participate in. Therefore, the psychological needs of relatedness are less significant in their sports endeavors.

4 Conclusion

In conclusion, gender has a significant moderating effect on the relationship between athletes' perception of coach leadership and athletes' basic psychological needs while the type of sports has a significant moderating effect on the relationship between athletes' basic psychological needs and sport motivation. This result revealed that the psychological needs in sports of athletes can be impacted through different types of leadership styles and that gender plays a role in the fulfillment of athletes' basic psychological needs by their coaches' leadership styles. Thus, coaches need to take gender into consideration when implementing leadership styles so that the psychological needs of athletes are been met by the different leadership styles of the coaches. Similarly, this study has shown that the type of sports does influence the relationship between the athletes' basic psychological needs and athletes' motivation. This finding shows that the different types of sports provide different avenues for athletes to fulfill their basic psychological needs, and hence, they are more motivated in their sports. Therefore, in their effort to motivate their athletes, coaches must take into consideration the different elements that can be offered by the different type of sports in fulfilling the athletes' psychological needs and consequently help to increase athletes' motivation in sports.

It is recommended future study should take into consideration the coaches' competency and level of experience in order to better understand the relationship the between constructs of interest.

References

1. Chu, R., & Tingzon, C. (2009). The relationship of coaching competency on the athlete's self-efficacy and hope. *The International Journal of Research and Review, 1,* 84–121.
2. Gillet, N., & Rosnet, E. (2008). Basic need satisfaction and motivation in sport. *The Online Journal of Sport Psychology, 10.*
3. Crust, L., & Lawrence, I. (2006). A review of leadership in sport: Implications for football management. *Athletic Insight: The Online Journal of Sport Psychology, 8,* 28–48.
4. Chelladurai, P., & Saleh, S. (1980). Dimensions of leader behavior in sports: Development of a leadership scale. *Journal of Sport Psychology, 2,* 34–45.
5. Vlachopoulos, S. P., Ntoumanis, N., & Smith, A. L. (2010). The basic psychological needs in exercise scale: Translation and evidence for cross-cultural validity. *International Journal of Sport and Exercise Psychology, 8,* 394–412.
6. Deci, E. L., & Vansteenkiste, M. (2004). Self-determination theory and basic need satisfaction: Understanding human development in positive psychology. *Ricerche di Psicologia.*
7. Hodge, K., Lonsdale, C., & Ng, J. Y. (2008). Burnout in elite rugby: Relationships with basic psychological needs fulfilment. *Journal of Sports Sciences, 26,* 835–844.
8. Bolkiah, S., & Terry, P. C. (2001). Coaching preferences of athletes in Brunei Darussalam and Great Britain: A cross-cultural test of the path-goal theory. In *Proceedings of the International Society of Sport Psychology 10th World Congress* (pp. 8–10).
9. Amorose, A. J., & Anderson-Butcher, D. (2007). Autonomy-supportive coaching and self-determined motivation in high school and college athletes: A test of self-determination theory. *Psychology of Sport and Exercise, 8,* 654–670.
10. Hollembeak, J., & Amorose, A. J. (2005). Perceived coaching behaviors and college athletes' intrinsic motivation: A test of self-determination theory. *Journal of Applied Sport Psychology, 17,* 20–36.
11. Ng, J. Y. Y., Lonsdale, C., & Hodge, K. (2011, June). The Basic Needs Satisfaction in Sport Scale (BNSSS): Instrument development and initial validity evidence. *Psychology of Sport and Exercise, 12,* 257–264.
12. Pelletier, L. G., Rocchi, M. A., Vallerand, R. J., Deci, E. L., & Ryan, R. M. (2013). Validation of the revised Sport Motivation Scale (SMS-II). *Psychology of Sport and Exercise, 14,* 329–341.
13. Hair, J. F., Jr., Hult, G. T. M., Ringle, C., & Sarstedt, M. (2013). *A primer on Partial Least Squares Structural Equation Modeling (PLS-SEM).* Sage Publications.
14. Beam, J. W., Serwatka, T. S., & Wilson, W. J. (2004). Preferred leadership of NCAA Division I and II intercollegiate student-athletes. *Journal of Sport Behavior, 27,* 3–17.

Relationship Between Agility and Power to Single-Sprint Performance and Repeated-Sprint Ability

Suhana Aiman, Sarina Md. Yusof, Zulkifli Abd Kadir and Ahmad Fuad Mohamad Nor

Abstract The objective of this study is to determine the relationship between agility and power to single-sprint performance and repeated-sprint ability. Thirty soccer players, aged between 19 and 25, participated in this study. Four tests were conducted: T-test, vertical jump test, 30 m sprint test and Bangsbo sprint test. The finding showed that there is a significant correlation between agility and single-sprint performance ($r = .46$, $p < .05$). A significant correlation also exists between agility and repeated-sprint ability ($r = .72$, $p < .05$). This study also found that there is a correlation between power and single-sprint performance ($r = -.70, p < .05$), and between power and repeated-sprint ability ($r = -.56, p < .05$). In conclusion, higher agility and power will result in better single-sprint performance and repeated-sprint ability.

Keywords Agility · Power · Single sprint · Repeated-sprint ability

1 Introduction

Success in soccer requires high-level in technical, tactical and physical skills. The rate of work of a football player ranges from low-level activities like walking, jogging to those of high-intensity such as sprinting [1]. Many instances in soccer demand rapid changes of direction [2]. In most team sports, directional changes are often initiated in response to some external stimuli such as movement of an opponent or ball [3]. Soccer requires high endurance, speed, agility and power at the elite level [1]. The external power is a determinant for acceleration and the maximal running velocity during sprinting [4].

In soccer, actions take place in a stochastic way; variations of intensity and directions are frequent and soccer players are required to perform these actions repeatedly and in different directions [5]. Players are often required to repeatedly produce maximal or near maximal sprints of short duration (1–7 s) with brief recovery periods.

S. Aiman (✉) · S. Md. Yusof · Z. Abd Kadir · A. F. Mohamad Nor
Faculty of Sports Science and Recreation, Universiti Teknologi MARA, Shah Alam, Malaysia
e-mail: suhana083@salam.uitm.edu.my

N. Sulaiman et al. (eds.), *Proceedings of the 3rd International Colloquium on Sports Science, Exercise, Engineering and Technology*,
https://doi.org/10.1007/978-981-10-6772-3_2

Therefore, the ability to repeat multiple sprints at high speed is important for soccer physical performance [6]. The aim of this study is to determine the relationship between agility and power to single-sprint performance and repeated-sprint ability among soccer players.

2 Method

2.1 *Participants*

A total of 30 male soccer players participated in this study. All participants are students of Universiti Teknologi MARA who were involved in the Inter-faculty Competition. Their age ranged between 19 and 25 years old. Injured players were excluded from the study.

2.2 *Testing*

The T-test was used to measure agility. Power was tested using vertical jump test. Single-sprint performance was assessed using 30 m sprint test. Repeated-sprint ability was evaluated using Bangsbo sprint test. Tests were conducted in 2 days. The players performed vertical jump test and T-test on the first day. On the following day, they performed 30 m sprint test and Bangsbo sprint test.

2.3 *Statistical Analysis*

The correlation between variables was determined by Pearson product-moment of correlation coefficient (r). The significant level for this study was set at $p < .05$. All analyses were performed using SPSS version 20.0.

3 Results and Discussion

3.1 *Results*

Table 1 shows the physical characteristics of the soccer players. The agility, power, single-sprint and repeated-sprint measurements of the soccer players in this study are displayed in Table 2.

Table 1 Physical characteristics of the soccer players

Variable	Mean ± SD
Age (years)	22.60 ± 0.86
Height (m)	1.70 ± 0.03
Body mass (kg)	66.75 ± 5.33
BMI (kg/m^2)	23.20 ± 1.73

Table 2 Test results for the soccer players

Variable	Mean ± SD
Agility (s)	10.98 ± 0.69
Power (cm)	45.13 ± 3.41
Single sprint (s)	4.52 ± 0.21
Repeated-sprint ability (s)	56.51 ± 2.61

Table 3 Correlation between agility, power, single-sprint performance and repeated-sprint ability

Variable	Agility	Power
Single sprint	.46*	−.70*
Repeated-sprint ability	.72*	−.56*

*$p < .05$

The correlation between variables is shown in Table 3. There was a moderate correlation between agility and single-sprint performance ($r = .46$, $p < .05$). A significant high correlation was found between agility and repeated-sprint ability ($r = .72$, $p < .05$). The power was highly correlated with single-sprint performance ($r = -.70$, $p < .05$) and repeated-sprint agility ($r = -.56$, $p < .05$).

3.2 Discussion

Finding from this study showed that there was a correlation between agility and single-sprint performance. This may due to the T-test protocol. The T-test involves a combination of leg speed, leg power and agility [7]. Of these three physical characteristics, leg speed had the highest partial correlation with the T-test [7]. It seems reasonable to assume that how fast an individual can sprint, shuffle or back-pedal would be determined primarily by leg speed [7].

The results also indicate that agility correlated with repeated-sprint ability. The capacity to perform short changes of direction has an important role in repeated-sprint ability [5, 8]. The Bangsbo sprint test was designed to involve sprinting in straight line and changing direction to mimic actual game condition.

In this study, a correlation was found between power and single-sprint performance. Powerful leg muscle is required to produce good sprinting [9, 10]. Muscle power is needed for acceleration and maintaining a maximal velocity in sprint performance [4]. The higher force production during sprinting is related to high power of lower body [11]. An increased in maximal power brings about in improved sprint ability [12]. Thus, an explosive power movement was found to be significant among the fastest players [13].

The current study also found that power correlated with repeated-sprint ability. The lower limbs power contributes to repeated-sprint ability [5]. This contribution is influenced by the great involvement of power in movements of deceleration, followed by change of direction and re-acceleration [5].

4 Conclusion

From the overall data analysis and research findings, it can be concluded that agility and power have significant relationship with single-sprint performance and repeated-sprint ability. Soccer players who have better agility and power would have better single-sprint performance and repeated-sprint ability.

References

1. Aguiar, M., Abrantes, C., Macas, V., Leite, N., Sampaio, J., & Ibanez, S. (2008). Effects of intermittent or continuous training for speed, jump and repeated-sprint ability in semi-professional soccer players. *The Open Sports Sciences Journal, 1,* 15–19.
2. Little, T., & William, A. G. (2005). Specificity of acceleration, maximum speed, and agility in professional soccer players. *Journal of Strength and Conditioning Research, 19*(1), 76–78.
3. Henry, G., Dawson, B., Lay, B., & Young, W. (2011). Validity of a reactive agility test for Australian football. *International Journal of Sports Physiology and Performance, 6,* 534–545.
4. Chelly, M. S., & Dennis, C. (2001, February). Leg power and hopping stiffness: Relationship with sprint running performance. *Medicine & Science in Sports & Exercise, 33*(2), 326–333.
5. Pasquarelli, B. N., Santos, A. L., Frisselli, A., Dourado, A. C., & Stanganelli, L. C. R. (2010). Relationship between the Bangsbo Sprint Test with sprint, agility, lower limb power and aerobic capacity tests in soccer players. *Revista Andaluza de Medicina del Deporte, 3*(3), 87–91.
6. Impellizzeri, M. F., Rampinini, E., Castagna, C., Bishop, D., Bravo, F. D., & Wisloff, U. (2008). Validity of a repeated-sprint test for football. *International Journal of Sports Medicine, 29*(11), 899–905.
7. Pauole, K., Madole, K., Garhammer, J., Lacourse, M., & Rozenek, R. (2000). Reliability and validity of the T-test as a measure of agility, leg power, and leg speed in college-aged men and women. *Journal of Strength and Conditioning Research, 14*(4), 443–450.
8. Wong, D. P., Chan, G. S., & Smith, A. W. (2012). Repeated-sprint and change-of-direction abilities in physically active individuals and soccer players: Training and testing implications. *Journal of Strength and Conditioning Research, 26*(9), 2324–2330.
9. Chelly, M., Cerif, N., Amar, M. B., Hermassi, S., Fathloun, M., Bouhlel, E., et al. (2010). Relationship of peak leg power, 1 maximal repetition half back squat, and leg muscle volume to 5-m sprint performance of junior soccer players. *Journal of Strength and Conditioning Research, 24*(1), 226–271.
10. Tonnessen, E., Shalfawi, S. A., Haugen, T., & Enoksen, E. (2011). The effect of 40-m repeted sprint training on maximum sprinting speed, repeated sprint speed endurance, vertical jump,

and aerobic capacity in young elite male soccer players. *Journal of Strength and Conditioning Research, 25*(9), 2364–2370.

11. Micheal, G. M., Jeremy, J. H., Mark, D. R., Chistopher, C., & Timothy, J. M. (2006). The effects of a 6 week plyometric training program on agility. *Journal of Sports Science and Medicine, 5*, 459–465.
12. Sharp, L. R., Troup, P. J., & Costill, L. D. (1982). Relationship between power and sprint freestyle swimming. *Medicine and Science in Sport and Exercise, 14*(1), 53–56.
13. Cronin, J. B., & Hansen, K. T. (2005). Strength and power predictors of sports speed. *Journal of Strength and Conditioning Research, 19*(2), 349–357.

Squash Backhand Stroke Analyses for Three Different Playing Levels in Malaysia

Diyana Zulaika Abdul Ghani, Zainal Abidin Zainuddin and Halijah Ibrahim

Abstract Our aim was to analyze backhand strokes performed among three different levels of playing in squash (professional elite, intermediate, and beginners). Thirty-five matches from three different tournaments, performed by 15 players divided equally for each group, were analyzed via hand notational analysis. Results indicated that the professional elite had recorded the greatest number of backhand shots [$n = 2382$ followed by the intermediate ($n = 2027$) and the beginners ($n = 1017$)]. Further analysis has shown that the frequencies were significantly differed between groups ($F = 8.39$, $p \leq 0.005$). The notational analyses have identified the drive as the most shots performed (BHDV) (40.9% of 5426 shots). Only three types of shots were frequently performed by all three groups which are BHDV (2220 of 5426 shots), the backhand crosscourt drive (BHXCDV) (853 shots), and the drop (BHDP) (595 shots). In addition, three backhand shots have marked significant difference between groups, which are BHDV ($F = 16.235$, $p \leq 0.000$), the backhand volley drive (BHVDV) ($F = 14.902$, $p \leq 0.001$), and BHXCDV ($F = 5.793$, $p \leq 0.017$). Even though different levels of playing execute different number of shots per game, it was evident that all three groups had performed same type of shots as their most frequently performed, and it was mainly at the back of the court (BHDV). These findings could give practical implication to assist coaches and players to enhance their playing strategy performance.

Keywords High-performance sport · Backhand analyses · Squash
Professional elite

D. Z. Abdul Ghani (✉) · Z. A. Zainuddin · H. Ibrahim
School of Education, Faculty of Social Sciences and Humanities, Universiti Teknologi Malaysia, Johor Bahru, Malaysia
e-mail: diyanazulaika@utm.my

N. Sulaiman et al. (eds.), *Proceedings of the 3rd International Colloquium on Sports Science, Exercise, Engineering and Technology*,
https://doi.org/10.1007/978-981-10-6772-3_3

1 Introduction

Performance in a squash depends on many factors, including tactical and technical strategies. A squash game consists of two major strokes during a play: the forehand and the backhand. In a real game, opponents always attack a player's weakest point mostly at the backhand area. The backhand is always a difficult stroke to execute, and it is the fragile part of most players [1] and least favored by players. Hence, players without strong backhand strokes are considered handicapped in squash.

A video on hand notational analysis conducted prior to this study for international tournament (National Sport Centre Series No. 6 one star in 2009) has shown that most of the elite players played almost 62% on the backhand side, while players with good backhand skill have the advantage during attacks and defensive situations. Such strokes have also been spotted in National Junior Circuit Sportexcel (2015) and National Junior Championships (2015). Most players had lost points following continuous attacks on their backhand area. Both situations clearly justify that backhand is a crucial factor influencing players' performance during the game.

The decision to execute strokes during the game is depending on the ball retrieved by the opponent which led to a variety of strokes. According to [2], players may use different types of strokes for different situations or same strokes for same situation; the player's choice is based on their observation of the opponent's game strategies. Different players may perform different playing styles according to their opponents' response shots. The stroke choice appeared to be different between levels of athletes; thus, the focus of this study is to analyze the frequency of strokes and types of backhand shots performed between professional elite, intermediate, and beginner players.

2 Method

In this study, "professional elite group" (PE) is defined as professional squash players with the highest competitive level according to national and international standards and also holding world rankings. The "intermediate group" (IM) are national junior squash players with strong knowledge of basic skills and currently under the National Sports Council's (NSC) development training program and Squash Racquets Association of Malaysia (SRAM) whereas The "beginner group" (BR) refers to squash players represent their university and only have basic knowledge of squash game.

2.1 Participants and Sample of Matches

A total of 15 Malaysian squash players (age 21.27 ± 5.22, height 1.69 ± 0.98 m, weight 64 ± 10 kg) (PE, $n = 5$; IM, $n = 5$ and BR, $n = 5$) were involved in this study.

The University ethics committee approved the ethical procedures and all participants signed the consent form.

Matches were recorded at three different tournaments: the CIMB Nicol David Kuala Lumpur Open for PE group, the National Junior Championship for IM group, and the Higher Educational Institutional Sports League for BR group. Thirty-five matches were recorded with a total of 35 matches [PE, ($n = 10$); IM, ($n = 15$); BR, ($n = 10$)]. Each player played a different number of games due to their opponent's performance.

2.2 *Procedure*

All matches were recorded using *Sony* Digital Video Camera Recorder, model DCR-PC350E (lens Carl Zeiss Vario-Sonnar T, 720×576 pixels, 25 frames per seconds), which was positioned at the court's frontal view. All recorded data were then analyzed and compared between groups using manual score sheets. Squash's coaches ($n = 2$) and experts ($n = 2$) were involved during notational analysis. An interobserver analysis was conducted to analyze the reliability of these observers. The coaches and experts had analyzed the types of shots from two intermediate group's matches, and a kappa coefficient (k) analysis was conducted. The kappa coefficient's agreement value of the types of shots from the two matches was acceptable ($k = 0.78$), and the observers were considered reliable to conduct further notational analysis.

3 Results and Discussion

3.1 *Types of Shots Performed*

Fifteen types of backhand shots were identified during the data analysis. They were serve (BHS), drive (BHDV), volley drive (BHVDV), crosscourt drive (BHX-CDV), volley crosscourt drive (BHVXCDV), boast (BHB), volley boast (BHVB), drop (BHDP), volley drop (BHVDP), crosscourt drop (BHXCDP), volley crosscourt drop (BHVXCDP), lob (BHLB), crosscourt lob (BHXCLB), volley crosscourt lob (BHVXCLB), and back mirror (BHBM).

Overall, findings showed that players from all three groups tend to play BHDV more often (2220 of 5426 shots) whereas BHXCDV was recorded as the second highest preferable shot (853 of 5426 shots). The third most commonly performed shot was BHDP (595 of 5426 shots) (Table 1).

All three groups have had recorded drive as their most frequent shots performed. The PE group had performed a total of 1003 backhand drive shots. The IM players have played a total of 976 shots while the BR have marked a total of 241 shots.

Table 1 Descriptive statistics for most frequent shots performed by three groups

Shots	Groups	Total	S error	Min	Max
BHDV	PE	1003	34.702	65	269
	IM	976	10.581	176	235
	BR	241	8.096	31	78
BHXCDV	PE	445	19.460	19	136
	IM	265	9.230	25	83
	BR	143	3.776	19	40
BHDP	PE	266	12.627	16	90
	IM	186	4.598	26	50
	BR	143	4.665	17	40

Table 2 Comparison between groups and the pattern of significance among those three types of shots

Type of shots/Groups	BHDV			BHVDV			BHXCDV		
	PE	IM	BR	PE	IM	BR	PE	IM	BR
PE		x	✓		✓	x		x	✓
IM	x		✓	✓		✓	x		x
BR	✓	✓		x	✓		✓	x	

✓ indicates significance between groups
x indicates no significance between groups

Among all 15 shots, only three were recorded significant difference between groups (Table 2). The shots were BHDV ($F = 16.235$, $p \leq 0.000$), BHVDV ($F = 14.902$, $p = 0.001$), and BHXCDV ($F = 5.793$, $p = 0.017$).

Table 2 explained the comparison between groups and the significance among three types of shots. Apparently, no exact pattern appeared between three groups regarding these three types of shots; however, there were still significant differences in terms of number of shots performed throughout game.

4 Discussion

This study found the high frequency of backhand shots played by PE (43.9%) compared to the other two groups which are IM (37.4%) and BR (18.7%). This evidence suggests that level of playing has influenced playing strategy. The differences in shots frequency were influenced by the tactical and technical skills performed by the players during tournaments. The difference in shot's frequency in a rally probably related to technical, tactical, and also the experiences of players. It is also might be influenced by the accuracy of shots execution during the rally which forced players to make extra movement to return the shots thus leads to more shots involved which correlate with previous findings [2, 3]. The PE players recorded highest backhand shots as they always involved in long rally during their game compared to the other groups which tend to quickly end their rally. This result confirms the study conducted by Hughes et al. [4] on comparing playing patterns between three different groups: elite, county, and recreational which resulted in different pattern encompasses by different standards of playing.

Playing strategies were developed through experiences and practices. This factor was evident in this study, which distinguishes playing standard between levels where higher level of players recorded higher number of shots. According to Ref. [5], individuals' intrinsic dynamics together with their set of knowledge may perform differently in various ways. Diverse movement develops while interacting between specific task and environment if a certain situation arises. In this case, the advantage is at the elite players side where they are able to continuously adapt to the demand of the games where changes of techniques and tactics are required. The adaptability might be due to more exposure to games and tournaments.

Interestingly, among the 15 types of shots, only three (BHDV, BHXCDV, and BHDP) were most frequently performed by the players. Consequently, the England Squash Federation had stated 56% of backhand drives (BHDV) post as winner distribution during matches. Reference [6] also found that winners and errors generally performed on the left side of the court and at the back of the court (backhand side). This concurs with the evidence of the higher number of shots frequency in BHDV.

Evidence also pointed that BHXCDV is one of most preferable shot played during games. Through crosscourt shots, players could change a ball's direction. This sudden change of direction could give a negative impact to the opponents as they have to return the hit, maintaining balance and making a turn simultaneously. This could lead to a poor return and create an opportunity to win a point. Reference [2] findings have supported the evidence of this study when their research concluded that players execute the highest percentage of strokes that are aimed to the back area of the court.

Findings had shown that BHDP has emerged as one of the frequently performed shots performed by players. Drop shots often force a player to run hastily toward the court's front area, which always disturbs the playing rhythm during the game. When perturbations happens, a player becomes unstable, which leads them making imprudent mistakes. Drop shots could increase an opponent's pressure by decreasing the time available for them to return the shot. This is in accordance with the previous study by Vucković et al. [7] who professed that players tend to play straight drop shots during attacks, when time constrains (less than 16 s). According to Refs. [6, 8], they stated that drop shots have high percentage of winner distributions.

From a strategy perspective, this high number of shots played at the back of a court is an indicator of a basic play as suggested in [2] and "playing safe" [7]. A drive shot acts as a strategic game shot as it places the return ball at the back of the court to ensure that the opponent is far from the optimum position in court (T-position); it also ensures that players had chances to plan another offensive shot to win a particular rally. Reference [8] has summarized shot frequencies during a game and concluded that the drive shots have obtained the highest frequency among others shots. This corroborated the findings of the present study. Thus, the drive shot is a basic shot with a high tendency to win points during a rally; hence, it is a shot that is frequently performed by players compared to other shots.

5 Conclusion

Evidently, there are no doubts drive shots (BHDV and BHXCDV) are the most frequent shots performed by all players in three groups; however, there are still significant differences between groups. This suggests that although types of shots are the same, the length of rally and the number of shots performed are in different levels. Probably, these factors differentiate their ranking in the squad. However, the findings of this paper verify that even though the players were from three different groups with different playing standard, they were able to produce the same type of shots as their favorite during a game.

This phenomenon could be related to the intrinsic dynamics of the dynamical system theory of movement [5]. Each individual has distinctive intrinsic dynamics and for novices, their prior intrinsic dynamics lead them to react immediately for effective performance without them realizing it. However, their intrinsic dynamics is in conflict state when they start to learn specific movements. Henceforth, this will differentiate their playing ability when they are becoming more matured and experienced. Findings of this study could be used by coaches to understand their players and to analyze tactical choices during the game. The findings also could be used as tools to search for the strengths and weaknesses of certain players. Eventually, players and coaches could formulate new strategies and training programs to enhance the performance for further achievements. Further research could use the same approach and analyze the winners and errors' shots produced by players during tournaments which were not analyzed in this study.

Acknowledgement This study was conducted using the Research University Grant (Q.J130000.7131.00J73) of Universiti Teknologi Malaysia (UTM), Malaysia.

References

1. Tolentino, B. (2009). Squash stroke execution mechanism.
2. Vučković, G., Dezman, B., Kovacic, S., & Pers, J. (2006). Quantitative analysis of playing efficiency in squash. In *IV Congreso Mundial de Ciencia y Deportes de Raqueta*.
3. Vučković, G., & James, N. (2010). The distance covered by winning and losing players in elite squash matches. *Kinesiologia Slovenica, 1,* 44–50.
4. Hughes, M., Well, J., & Matthews, K. (2000). Performance profiles at recreational, county and elite levels of women's squash. *Journal of Human Movement Studies, 39,* 85–104.
5. Davids, K., Button, C., & Bennett, S. (2008). *Dynamics of skill acquisition. A constraints-led approach*. Human Kinetics.

6. Hughes, M., Howells, M., & Hughes, M. (2006). Using perturbations in elite men's squash to generate performance profiles. In A. Rendimiento (Ed.), *IV Congreso Mundial de Ciencia y Deportes de Raqueta* (pp. 35–51).
7. Vucković, G., James, N., Hughes, M., Murray, S., Sporis, G., & Pers, J. (2013). The effect of court location and available time on the tactical shot selection of elite squash players. *Journal of Sports Science & Medicine, 12,* 66–73.
8. Murray, S., & Hughes, M. (2001). *Tactical performance profiling in elite level senior squash.* Wiltshire: Crowood Press.

Effects of Concurrent, Strength, and Endurance Training on Metabolic Syndrome Risk Factors Among Obese Females

Sarina Md. Yusof, Noor Izzati Mohd Idris, Suhana Aiman and Zulkifli Abd Kadir

Abstract The purpose of this study was to investigate the effects of a concurrent, strength, and endurance training program on metabolic syndrome risk factors among obese females. Fifty-two obese females (mean ± SD: age 22.71 ± 1.53 years; body mass index 33.96 ± 3.36 kg m^{-2}) were randomly assigned into four groups ($n = 13$; concurrent (CE), endurance exercise (EE), resistance exercise (RE) programs, and control (C). All the intervention groups completed an 8-week intervention period that consisted of either an endurance exercise (EE: aerobic exercise intensity level started at 40% of training heart rate at week 1 and increased to 60% of training heart rate at week 8), resistance exercise (six to eight resistance exercises using machine at 50–70% of predicted 1RM), or concurrent exercise (CE: 25-min running plus six to eight resistance exercises at 50–70% of predicted 1RM). Resistance exercise was focused on upper body, lower body, and abdominal muscles. Before and after the intervention, subjects completed an incremental treadmill run and maximal isometric strength tests. The baseline values for waist circumference, TRI, HDL, fasting blood glucose, and blood pressure were identical in all groups. Significant interaction (time x intervention) were found in WC, systolic blood pressure, and blood glucose, but not in TRI and HDL. However, no significant differences were found between the groups in all outcome measures. The results suggest no significant difference in benefits of an 8-week of endurance, resistance and concurrent exercises on WC, TRI, HDL, fasting blood glucose, and blood pressure among obese female adults maybe because of an insufficient sample size or a short intervention period.

Keywords Concurrent · Strength · Endurance · Metabolic syndrome

S. Md. Yusof (✉) · N. I. Mohd Idris · S. Aiman · Z. Abd Kadir
Faculty of Sport Science and Recreation, Universiti Teknologi MARA, Shah Alam, Selangor, Malaysia
e-mail: sarin864@salam.uitm.edu.my

N. Sulaiman et al. (eds.), *Proceedings of the 3rd International Colloquium on Sports Science, Exercise, Engineering and Technology*,
https://doi.org/10.1007/978-981-10-6772-3_4

1 Introduction

Obesity is generally associated with excessive body weight which is caused from excessive fat in the body due to an imbalance between energy intake and energy expenditure. Metabolic syndrome (MetS) is a group of risk factors which increases the heart disease and health problems. The five risk factors are large waist circumference (WC), high triglyceride (TRI) and low high-density lipoprotein (HDL) levels, high fasting blood sugar, and blood pressure (BP). Metabolic syndrome is becoming very common due to an increase in obesity rate in adults. Having metabolic syndrome will increase the risk of developing cardiovascular disease and diabetes. Presence of 3–5 risk factors constitutes the diagnosis of metabolic syndrome [1]. Malaysia has been noted to have a high prevalence of MetS when compared to other Asian countries [2].

Prevalence of metabolic syndrome are more evident in women as compared to men [3]. Evidences have shown that in most cases of obesity are more related to low energy expenditure. Weight training or resistance exercise shows many advantages, which is a powerful stimulus to increase mass, strength, muscular power and helps to preserve the integrity the musculature.

Literatures have shown that resistance training is a safe training method and significantly help to improve body weight, reducing waist circumference, increasing muscular strength and improve physical performance [4, 5]. Endurance exercise helps reduce cardiovascular risk by reducing body weight, improving blood lipid profiles [6]. There is evidence to show that the combined resistance exercise and aerobic activity improve blood lipid profiles better than either exercise performed independently [7]. However, there are evidences showing that combining the exercises may compromise the gains from weight exercise as compared to when conducting it alone due to the interference effect. Concurrent exercise is defined as the integration of endurance and weight training in a periodized exercise regime [8]. The purpose of this study was to evaluate the effect of endurance, resistance, and concurrent training on metabolic risk factors among obese female adults.

2 Method

2.1 Participants

A total of 52 participants who met the inclusion criteria were recruited via advertisement. Participants were adults obese female volunteers aged between 20 and 26 years old and underwent two screening processes. Criteria for inclusion were: females having Body Mass Index (BMI) ranged from 30.0 to 39.9 kg/m^2, body fat percentage ranged between 35 and 45%, and free from any chronic diseases (diabetes mellitus, hypertension, any cancer disease, and any of heart diseases and endocrine disorder). Volunteers with musculoskeletal injuries, physically impaired, under medication or drugs or both, and having abnormal menstruation were excluded. Participants were required to pass a Physical Activity Readiness Questionnaire (PAR-Q), health status examination and ECG tests prior to participation. Participants gave their consent upon arriving at the lab. The study was approved by UiTM Ethics Committee in accordance to the Helsinki Declaration.

Participants were randomly assigned based on age, BMI, body fat percentage, and non-fasting blood glucose into four groups: aerobic exercise (AE), resistance exercise (RE), concurrent exercise (CE), and control. Participants' characteristics are as shown in Table 4. All participants successfully completed the intervention, and no injuries due to the intervention were reported. Outcome measures were measured for the pre- and post-intervention.

2.2 Intervention

All exercise groups engaged with the exercise program for 8 weeks, 3 times per week, and 1 h per session. Participants warmed up for 5 min prior to exercise and cooled down for 5 min after the conditioning phase. Earlier, participants were taught to perform carotid palpation to measure their resting heart rate upon wake up from sleep in the morning. Carotid palpation was used to calculate the training heart rate for aerobic exercise group.

Aerobic exercise (AE)—The prescribed aerobic exercise intensity level started at 40% of the training heart rate at week 1 and increased to 60% of training heart rate at week 8 of the intervention. Aerobic exercise phase began with a brisk walk to slow jog (Table 1).

Resistance exercise (RE)—Resistance exercise group were asked to perform six to eight resistance exercises using machine at 50–70% of predicted one repetition maximum (1-RM) with rest interval between 1 and 2 min between set [9]. The exercises were focused on upper body, lower body and abdominal muscles (Table 2).

Concurrent exercise (CE)—For concurrent exercise group, the participants had to perform aerobic exercise for 25 min followed by resistance exercise in one session at the same intensity with aerobic and resistance exercise. Participants in concurrent group were trained half of the volume of resistance exercise (Table 3).

Control

The control group did not follow any exercise program but maintained a normal daily activity throughout the exercise duration.

Table 1 Aerobic exercise program

Exercise protocol	Exercise routine
Intensity: 40–60% of heart rate maximum (HR_{max})	Treadmill brisk walking Treadmill jogging

Table 2 Resistance exercise program

Exercise protocol	Exercise division	Exercise routine
Intensity: 50–70% of 1RM	Two upper body exercise	Chest press Lat pull down Shoulder press Biceps curl
	Two lower body exercise	Leg extension Leg curl Leg press Calf raise
	Two abdominal exercise	Russian twist Abs crunch Leg raise Plank

Table 3 Concurrent exercise program

Exercise method	Exercise protocol	Exercise routine
Combining aerobic and resistance exercise in one session	Resistance training Intensity: 50–70% of 1RM Duration: 30 min Frequency: 3× per week Exercise: 6–8 exercises	Similar to exercise routines implemented in resistance exercise
	Aerobic training Intensity: 40–60% of HR_{max} Duration: 30 min Frequency: 3× per week	

Participants were advised to maintain to their normal diet throughout the exercise program. Three days food record was collected from the participants to ensure they were keeping to the normal diet.

2.3 *Outcome Measures*

Anthropometry measurements

Height and weight were measured using stadiometer and weighing machine (Tanita). Body mass index is calculated using mass formula (kg) divided by height × height (m^2).

Waist circumference

WC was measured at the end of several consecutive natural breaths, at the level parallel to the floor, midpoint between the top of the iliac crest and the lower margin of the last palpable rib in midaxillary line [10]. The data were analyzed using cutoffs points for Caucasians and Asians (80 cm in women) [11].

Triglyceride and high-density lipoprotein

Venous blood samples were taken by a doctor and analyzed for lipid profile and fasting glucose. Triglycerides were measured based on an enzymatic procedure. The Abell–Kendall method was used to measure high-density lipoprotein.

Fasting blood sugar

Fasting serum glucose was assayed enzymatically by using the VISTA kit analyzer (SIEMENS, USA). The process involves in sampling the blood specimens, reagent delivery, mixing procedures, and chemical processing were automatically performed by Dimension Vista software operating system (2008).

Blood pressure

This systolic and diastolic blood pressure measurement was using manual desk mercury sphygmomanometer. The measurement was taken at the left arm after participants sit at backrest chair for 10 min. The values of systolic and diastolic were recorded in mmHg. Reliability for systolic was reported at $r = .89$ and diastolic was reported at $r = .83$ [12].

2.4 Data Analysis

This study utilized a mixed within-between subjects analysis of variance (ANOVA) to determine the significant interaction and main effects of the interventions on the outcome measures between pre- and post-intervention. Data are reported as mean ± SD. The statistical significance level was set at 0.05. Statistical Package for Social Science version 21.0 was used to analyze the data. Tukey's post hoc test was further used to determine the differences.

3 Results and Discussion

3.1 Participants' Characteristics

Participants were not significantly different in age, weight, height, WC, TRI, HDL, and blood glucose at baseline. Table 4 shows the participants' characteristics.

3.2 Effects of Intervention on WC

Significant interaction between intervention type and time was found for WC, Wilks' Lambda = .597, F (3, 48) = 10.81, $p < .05$, partial eta squared = .043, with all groups

Table 4 Participants' baseline characteristics

	AE ($n = 13$)	RE ($n = 13$)	CE ($n = 13$)	Control ($n = 13$)
Age (years)	22.38 (.768)	22.46 (1.266)	23.77 (1.787)	22.23 (1.691)
Weight (kg)	83.34 (9.59)	86.81 (12.71)	87.78 (10.27)	86.19 (9.78)
Height (cm)	157.31 (4.12)	159.72 (5.84)	160.59 (4.81)	158.82 (5.35)
BMI (kg/m^2)	33.62 (3.26)	34.01 (4.09)	34.04 (2.99)	34.18 (3.38)
WC (cm)	87.79 (6.33)	91.50 (7.29)	91.72 (8.58)	91.06 (7.87)
TRI (mmol/L)	1.23 (.56)	.96 (.22)	1.02 (.42)	.95 (.46)
HDL (mmol/L)	1.25 (.26)	1.37 (.26)	1.28 (.29)	1.33 (.25)
Blood glucose (mmol/L)	4.90 (.77)	4.57 (.44)	4.90 (.47)	4.70 (.56)
Systolic pressure (mm Hg)	124.31 (12.15)	122.62 (12.85)	124.38 (11.97)	127.08 (10.24)
Diastolic pressure (mm Hg)	76.77 (11.83)	77.15 (9.15)	78.77 (10.51)	76.92 (6.93)

showing reduction in WC. The main effect for time, Wilks' Lamda = .615, F (1, 48) = 30.02, $p < .05$, partial eta squared = .385, with three groups showing reduction in WC across two observations. The main effect comparing the types of training was not significant, F (1, 48) = 1.669, $p = .186$, partial eta squared = .094, suggesting no difference in the effectiveness of the three interventions.

This study supported earlier findings [13] which indicate that all types of exercise induce positive changes to WC, in which high WC measurement is a risk factor of metabolic syndrome which leads to cardiovascular disease [14]. Most of the studies which reported the changes were mostly conducted at a longer duration (12 weeks), while in the present study, it was conducted only in 8 weeks.

3.3 *Effects of Intervention on TRI and HDL*

No significant interaction between intervention (aerobic, resistance, concurrent, and control) and time on TRI, Wilks' Lambda = .957, F (3, 48) = 1.481, $p = .231$, partial eta squared = .085. There was no main effect for time, Wilks' Lambda = .925, F (1, 48) = 3.893, $p = .054$, partial eta squared = .075. The main effect when comparing the types of training also showed no significant difference, F (3, 48) = .714, p = .548, partial eta squared = .048, which suggests no difference in the effectiveness of all interventions in triglycerides.

A significant interaction between interventions (aerobic, resistance, concurrent, and control) and time in HDL is noted, Wilks' Lambda = .844, F (3, 48) = 2.956, p = .042, partial eta squared = .156. There was no main effect for time, Wilks' Lambda

= .975, F (1, 48) = 1.236, p = .272, partial eta squared = .025. The main effect when comparing the types of training showed no significant difference, F (3, 48) = .644, p = .591, partial eta squared = .039, which suggests no difference in the effectiveness of all interventions in high density lipoprotein.

Although some studies have shown improvement in TRI and HDL following 8 weeks of concurrent and endurance training [15], this study however did not show any improvement in TRI and HDL, but changes in body weight and fat mass were found. This is probably due to the difference in the exercise intensity and volume which did not exert any changes in the activities of some lipid-regulating enzymes [16]. For future study, it is suggested that to determine the adequate volume and intensity, cortisol level is to be measured.

3.4 *Effects of Intervention on Fasting Blood Glucose*

No significant interaction were found between training types and time in blood glucose, Wilks Lamda = .854, F (3, 48) = 2.739, p = .054, partial eta squared = .146. A substantial main effect for time, Wilks Lamda = .888, F (1, 48) = 6.051, p = .018, partial eta squared = .112, with all intervention groups shows reduction in blood glucose. The main effect when comparing the three interventions showed no significant difference, F (3, 48) = .865, p = .466, partial eta squared = .051, which indicates no difference in the effectiveness of all interventions in blood glucose level.

This study supported earlier study [16] which indicated improvement in blood glucose among obese nondiabetic subjects following endurance and concurrent training. This finding is further supported by improvement in WC. Improvement in blood glucose is associated with improved in insulin sensitivity [16].

3.5 *Effects of Intervention on Resting Blood Pressure*

Significant interaction between interventions and time were noted, Wilks' Lambda = .851, F (3, 48) = 2.805, p = .050, partial eta squared = .149. There was substantial main effect for time, Wilks' Lambda = .756, F (1, 48) = 15.487, p = .000, partial eta squared = .244. The aerobic, resistance, and concurrent groups showed a reduction in systolic blood pressure, but an increment of systolic blood pressure is shown in the control group. However, the main effect when comparing aerobic, resistance, concurrent, and control groups showed no significant difference, F (3, 48) = 1.821, p = .156, partial eta squared = .102, which suggests that there is no difference in the effective of all interventions in reducing systolic blood pressure.

Several studies have shown the improvement in systolic blood following endurance, resistance, and concurrent training [17]. The reduction in systolic blood pressure is probably due to the reduction in peripheral vascular resistance [18],

Table 5 Effect of intervention on WC, TRI, HDL, BG, and BP

	AE ($n = 13$)	RE ($n = 13$)	CE ($n = 13$)	Control ($n = 13$)
WC (cm)	84.18 (5.62)*	87.43 (5.46)*	91.20 (8.76)*	91.64 (7.26)
TRI (mmol/L)	1.05 (.58)	.92 (.21)	.90 (.34)	.99 (.40)
HDL (mmol/L)	1.23 (.20)	1.38 (.30)	1.41 (.32)	1.30 (0.26)
Blood glucose (mmol/L)	4.66 (.90)	4.36 (.39)	4.33 (1.06)	4.85 (.44)
Systolic pressure (mm Hg)	116.31 (8.60)	117.69 (11.66)	116.54 (7.47)	128.08 (9.65)
Diastolic pressure (mm Hg)	74.92 (7.64)	73.85 (9.61)	76.00 (8.64)	78.23 (6.83)

alteration of sympathetic nervous system function and vasculature responsiveness, and baroreflex action in lowering the blood pressure [18].

No significant interaction between and time on diastolic pressure was found, Pillai's Trace = .028, F (3, 48) = .460, $p = .711$, partial eta squared = .028. No main effect for time, Pillai's Trace = .024, F (1, 48) = 1.181, $p = .283$, partial eta squared =.024. In addition, the main effect when comparing aerobic, resistance, concurrent, and control groups showed no significant difference, F (3, 48) = .281, $p = .839$, partial eta squared = .017, which suggests that there is no difference in the effectiveness of all interventions in diastolic blood pressure (Table 5).

4 Conclusion

It can be concluded that these exercise interventions do not show clear superiority and definite adaptations in metabolic risk factors among obese female adults, however, it is proven that exercise intervention do induce positive changes in improving the metabolic risk factors. Further studies are, therefore, necessary to which include a larger sample size, different gender, and a longer intervention period.

Acknowledgements The authors would like to express gratitude to all participants participated in this study and staff of Faculty of Sports Science and Recreation, Universiti Teknologi MARA Shah Alam, Selangor for their support.

References

1. Alberti, K. G. M. M., Eckel, R. H., Grundy, S. M., Zimmet, P. Z., Cleeman, J. I., Donato, K. A., et al. (2009). Harmonizing the metabolic syndrome: A joint interim statement of the International Diabetes Federation Taskforce on Epidemiology and Prevention; National Heart, Lung, and Blood Institute; American Heart Association; World Heart Federation; International Atherosclerosis Society; and International Association for the Study of Obesity.
2. Beigh, S. H., & Jain, S. (2012, July). Prevalence of metabolic syndrome and gender differences. *Bioinformation, 8*(13), 613–616.
3. Mohamud, W. N., Ismail, A. A., Sharifuddin, A., Ismail, I. S., Musa, K. I., Kadir, N. A., et al. (2011, February). Prevalence of metabolic syndrome and its risk factors in adult Malaysians: Results of a nationwide survey. *Diabetes Research and Clinical Practice, 9*(2), 239–245.
4. Jan Mohamed, H. J., Mitra, A. K., Zainuddin, L. R., Leng, S. K., & Wan Muda, W. M. (2013). Women are at a higher risk of metabolic syndrome in rural Malaysia. *Women and Health, 53*(4), 335–348.
5. Mekary, R. A., Grøntved, A., Despres, J.-P., De Moura, L. P., Asgarzadeh, M., Willett, W. C., et al. (2015). Weight training, aerobic physical activities, and long-term waist circumference change in men. *Obesity, 23,* 461–467.
6. Mann, S., Beedie, C., & Jimeniz, A. (2014). Differential effects of aerobic exercise, resistance training and combined exercise modalities on cholesterol and the lipid profile: Review, synthesis and recommendations. *Sports Medicine, 44,* 211–221.
7. Wescott, W. L. (2012). Resistance training is medicine: Effects of strength training on health. *Current Sport Medicine Report, 11*(4), 209–216.
8. Pitsavos, C., Panagiotakos, D. B., Tambalis, K. D., et al. (2009). Resistance exercise plus aerobic activities is associated with better lipids profile among healthy individuals: The ATTIICA study. *QJM, 102,* 609–616.
9. Fyfe, J. J., Bishop, D. J., & Stepto, N. K. (2014). Interference between concurrent resistance and endurance exercise: Molecular bases and the role of individual training variables. *Sports Medicine, 44*(6), 743–762.
10. Kraemer, W. J., & Ratamess, N. A. (2004). Fundamentals of resistance training: Progression and exercise prescription. *Medicine and Science in Sports and Exercise, 36*(4), 674–688.
11. World Health Organization (WHO). (2008). *Waist circumference and waist-hip ratio.* Report of WHO Expert Consultation. Geneva: World Health Organization.
12. Beam, W., & Adams, G. (2008). *Exercise physiology manual* (7th ed.).
13. Norfazilah Ahmad, A., Samial, I. M. A., Azmawati, M. N., Mohd Rohaizat, H., & Hasanain Faisal, G. (2016). Abdominal obesity indicators: Waist circumference or waist-to-hip ratio in Malaysian adults population. *International Journal of Preventive Medicine, 7,* 82.
14. Adegbija, O., Hoy, W., & Wang, Z. (2015). Prediction of cardiovascular disease risk using waist circumference among Aboriginals in a remote Australian community. *BMC Public Health, 15,* 57.
15. Ali-Mohamadi, M., Abbaspoor, M., Rahimi, R., & Hakimi, M. (2014). The influence of order execution components of the strength and endurance in the concurrent training on lipid profile and body composition in overweight females. *World Applied Sciences Journal, 29*(7), 946–953.
16. da Silva Medeiros, N., de Abreu, F. G., Colato, A. S., de Lemos, L. S., Ramis, T. R., Dorneles, G. P., et al. (2015). Effects of concurrent training on oxidative stress and insulin resistance in obese individuals. *Oxidative Medicine and Cellular Longevity, 2015.*
17. Dias, I., Farinati, P., De Souza, M. G., Manhanini, D. P., Balthazar, E., Dantas, D. L., et al. (2015, December). Effects of resistance training on obese adolescents. *Medicine and Science in Sports and Exercise, 47*(12), 2636–2644.

18. Delavar, S. H., & Faraji, H. (2011). Effect of different concurrent training methods on post-exercise hypotension in borderline hypertensive women. *Middle-East Journal of Scientific Research, 9*(4), 456–461.

A Systematic Review of Type of Injury Among Rugby Union Players

Megat Ahmad Aslam Megat Azman, Norazhan Che Lan, Siti Hartini Azmi and Norasrudin Sulaiman

Abstract Rugby union is a full-contact sport that the rate of injuries is higher. Multiple studies discovered rate and type of injuries occurred in rugby union influences by many factors. Hence, the aim of this study is to identify the type of injury occur in rugby union and identify the comparison of approach on leading to injury. In addition, the study is also able to ascertain injury rate based on player position. Therefore, the study was conducted by systemic review on previous articles and journals, then the data will analyze by meta-analysis method. The result shown probability of injury occurs as high as 57.2% per 1000 player-hours. The results also showed that most of the injuries were soft muscle injuries and position of player did not influence the injuries rate and type.

Keywords Rugby union · Injury

1 Introduction

International Rugby Board (IRB) started to introduce professionalism in October 1995 Rugby World Cup 1995, where the player started to register as a professional player. The governing body IRB started to provide a financial reward to a professional rugby player in order to sustain the game quality. However, Ref. [1] states that due to professionalism involvement, the rugby players need to meet the physical and mental standard as well as show the pace and strength expected of a full-time athlete. Nevertheless, the Ref. [2] state that a standard expectation has also affected the majority of amateur players. Hence, the injury pattern and rate are becoming difficult

M. A. A. Megat Azman · N. Che Lan · N. Sulaiman (✉)
Faculty of Sport Science and Recreation, Universiti Technologi Mara (UiTM Shah Alam), Shah Alam, Malaysia
e-mail: noras878@salam.uitm.edu.my

S. Hartini Azmi
Faculty of Sport Science and Coaching, Universiti Pendidikan Sultan Idris, Tanjung Malim, Perak, Malaysia

N. Sulaiman et al. (eds.), *Proceedings of the 3rd International Colloquium on Sports Science, Exercise, Engineering and Technology*,
https://doi.org/10.1007/978-981-10-6772-3_5

to control among the players. According to Ref. [3], rugby resulted in the highest injury in sport where (198 cases/10,000 h) it is nearly double as basketball injury rate (103 cases/10,000 h). Nevertheless, updated that rugby still at the highest rate of injury followed by soccer and field hockey [4].

A different type of injury coming from different approaches of body contact in a rugby game. The damage can lead from mild to severe injury, categorized as foul play, tackle, scrum, and many more. The tackle has been reported in many articles that the action of the play that resulting in most injuries [5]. However, some research reported to have a high incidence of foul play. In addition, physical factor such as fatigueness has been suggested as one of the main influences to a rugby injury and only happen within a time of the game. The researcher believes that more injuries should occur during the second half of a match when players are fatigued [6].

Therefore, identifying the type of injury and comparing the different type of approach leading to injury in rugby union stand to be advanced via a reviewing number of articles and journals in meta-analysis of eligible studies. The aim of this systematic review is to examine the brief idea of the different type of injury in rugby union. This analysis will be limited to prospective studies only. Our main objectives are to determine the comparison of different approaches leading to injury and reviewing the most dangerous approach in rugby union that supported by past articles.

2 Method

2.1 Research Design

This research design is a systematic review, journals related to injury in rugby union were reviewed. The data collected and look from multiple studies. In addition, resources are mostly reviewing the specific demand samples and subjects to produce the topic requirement. The journals or sources are searching and analyze from the related question. The methodology part of the review will be listed in the databases and citation list. Examples of search database used were EMBASE, SPORTDiscus, PubMed, and Web of Sciences. In this research design, there are no participant and treatment involved. The brief summary of the type of injury coming from articles that revising the same sport (rugby union), where participants were controlled and have been properly briefed and tested. The related articles will provide the type of injury and approach that lead to injury, which will be the main focus of this journal review. Hence, this research design will help the journal review to generate the new idea and issue that are related to the future.

2.2 *Sample*

The samples are collected from previous journals and articles, then it will be extracted and analyzed for the review. This research sample will be utilizing the online search engine PubMed, EMBASE, and SPORTDiscus in October 2015. We then hand-researched the references from the related topic from key systemic reviews and prior study on rugby union.

2.3 *Data Collection*

The data collection for this journal is coming from different sources (direct.com, journal sport science and medical.com and many more), where the liability of the sources is very liable and established. The strategy during the journal searching in PubMed and adapted it for EMBASE: ((rugby union[tiab]) AND ((injury OR issue OR problem OR accident OR concussion OR bruise OR dislocation OR fracture OR sprain OR strain OR contusion OR hematoma OR laceration OR broken)) AND ((type OR mechanism OR tackle OR scrum OR approach OR professional OR amateur).

The documents are mostly downloaded in PDF format and then it will be categorized into different categories. The journal selection initially searches by title and the abstract reviewed looking for rugby injury and related field. Then they studied and marked as potentially beneficial to the study. The journal must contain the valuable information fitting the criteria. The information and text were then retrieved electronically online. The inclusion was primary search on rugby with an incidence on the type injury and mechanism of injury.

2.4 *Reliability and Validity*

Of the 500 studies found through online database, we only selected 30 which are included in the final review. The procedure of exclusion is given in the flow diagram given (Refer Fig. 1).

3 Results and Discussion

Studies recounted match injury rates as per 1000 player-hours. The probability percentage provided by six studies reported that injuries might occur as high as 57.2% per 1000 player-hours. The average probability of injury for one player can get up to 68%, where minimal probability can get up to 6% (Refer Table 1).

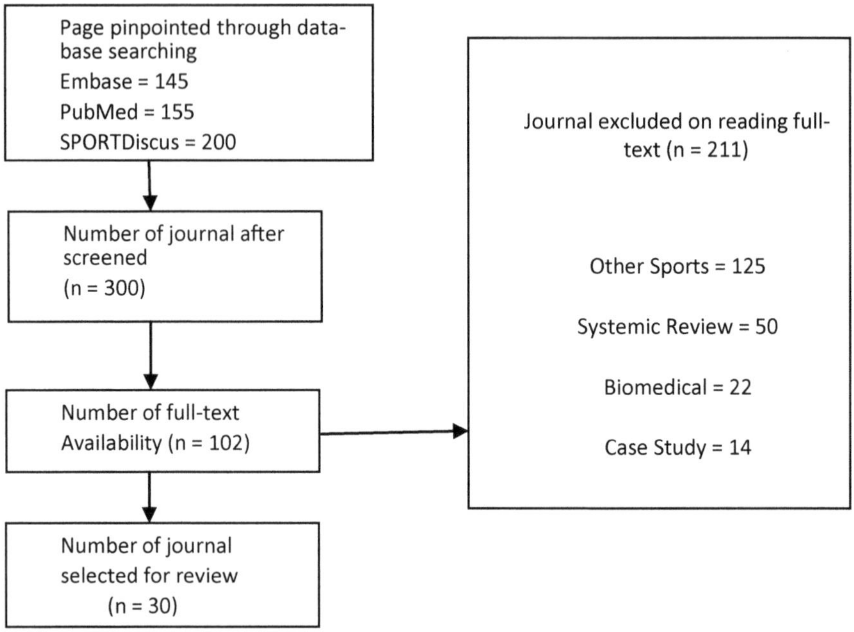

Fig. 1 Flowchart of literature selection

3.1 *Type of Injury*

Soft tissue and closed injuries accounted for almost half percent of all injuries. These were further subdivided into contusion/hematoma (46%) and joint/ligament sprains/tears (47.2%). Other types of injury included fractures (27%), dislocations/subluxations (10.8%), and concussions (24.6%) (Table 2).

3.2 *Phases of Play*

It is reported that phases of play (mechanism) play an important factor leading to different types of injury. In total, tackle, which contain tackling and being tackled reported as the majority rate in injuries (39.6–64%). It is measured that active tackling (16.5–65.0%) consisted of higher rate than active tackling (18.5–40%). It is followed by ruck and maul with injuries rate (8.3–31.5%) after tackling and being tackled. Lastly, injuries rate produced by scrum (2.0–36.0%) is the last phases of play in rugby union (Table 3).

Table 1 Probability of injury in percentage

Author and year of publication	Number of injuries	Total time exposure	Injury incidence (95% Cl were given)	Averange probability of a player getting injured
Fuller (2011)	190	3320 player-hour	57.2 (49.6–66.0) injuries per 1000 player-hours	63
Gabbett (2008)	62	1092 player-hour	56.8 (42.6–70.9) injuries per 1000 player-hours	68
Haseler (2010)	210	1636 player-hour	11.9 (4.1–19.6) injuries per 1000 player-hours	9
Nathan (1983)	10	2700 player-hour	3.7 injuries per 1000 player-hours	6
Nicol (2011)	26	2406 player-hour	10.8 injuries per 1000 player-hours	10
Roux (1987)	353	50 126 player-hour	7.0 injuries per 1000 player-hours	12

[a]All the information provided in their research itself (6 studies)

Table 2 Percentage of type of Injury

Injury	Percentage of all injuries (%)	Number of studies
Fracture	3.0–27.0	9
Ligament injuries, sprains and strains	15.7–47.2	7
Dislocation and subluxation	0.5–10.8	6
Laceration, contusion and haematoma	2.7–46.0	7
Concussion	2.2–24.6	10

[a]Data provided by 10 studies

Table 3 Percentage of phases of play that lead to injury

Phase of play	Percentage of all injuries	Number of studies
All tackle	39.6–64.0	9
Active tackle	18.5–40.0	8
Recipient of tackle	16.5–65.0	8
Scrum	2.0–36.0	9
Ruck/maul	8.3–31.5	7

[a]Data provided by 9 studies

Table 4 Injury percentage by player

Player position	Percentage of all injuries
Forwards	43.8–56.3
Backs	43.6–56.3

[a]Data provided by 9 studies

3.3 *Player Position*

Based on 13 studies reported that the data provide in two different types of positions (forward and back) in rugby union that lead to the injury are measurable (refer Table 4). The injury rate provided showed that the injuries occurred between two different positions 43.8–56.3% and 43.–56.35% of all injuries are slightly different.

Based on 13 studies reported that the data provide in two different type of positions (forward and back) in rugby union that lead to the injury are measurable (refer Table 4). The injuries rate provided showed that the injuries occurred between two different positions 43.8–56.3% and 43.6–56.35% of all injuries are slightly different.

3.4 *Site of Injury*

The highest possible parts where injuries may occur are lower extremity compared to the upper extremity. Based on 5 studies, injury occurs in upper limb and lower limb, which ranged from 19.3–38.4% and started from 3.4–46.8%.

4 Discussion

This study focuses on brief overview on the type of injury in rugby union and comparison on different approaches that leads to injury. Rugby has been well known full contact sport yet wear a little or no protective gear. Due to term professional involves the financial reward are given to the professional rugby player. Therefore, they need to achieve a standard strength and performance in training and competition. Reference [1] states that due to the high expectation, injury rate become uncontrolled. This is because the amateurs are very limited in experience where it leads to injury. Relevance to the percentage of probability for a player to get an injury is more than half (68) percent per 1000 player-hours. Therefore, the grade of professionalism in sport does play a significant role in many things especially the injury rate when it comes to rugby union [6].

Injuries in rugby have become very familiar with its own nature of the game, where body contact is vital to win a match. Reference [6] mention that the reason why injury in rugby union is higher due to the speed of the game, player are bigger and fitter, tackling is harder and tactical are complicated. The mechanism of injury is very complicated, and the vigorous impact during the game and training depends on the area being hit [3]. Majority of injury resulted from contact phases of play, where the studies have proven almost 65% of injury resulted from being tackled. However, the position of the player does not influence the injury percentage based on this systemic review. Therefore, understanding the phases of play that lead to injury may identify the amount of damage physically to the player and type of injury.

Previous research describes that the most percentage of the injuries located at the lower part of the body, 25% of injuries in knee consider severe [6]. Which supported by previous studies where ligament injuries, sprains and strains have the highest injury chances (47%). Although more severe injuries such as fractures and concussions are commonly reported [7]. The approaches in rugby such as scrums and tackle above shoulder may lead to upper body damages and severe medical injuries. Furthermore, the head is the most sensitive part of the body where the damages in rugby are very common [2]. In addition, the injuries resulted from scrums and bad tackle may also lead to serious neck injury [7]. Based on five studies, injury occur in upper limb and lower limb, ranging from 19.3–38.4% and started from 3.4–46.8%. Therefore, the phases of play in rugby can lead to upper limb injuries and possible severe medical issue.

5 Conclusion

It is theoretically accepted that rugby union is a physical sport and contributes to physical injuries. However, it is also associated with higher injury risk when it comes to rugby union and professionalism. Rugby is the physical sport that causes various types of injury where ligament, sprain, and strain are considered the highest injury ratio. It is believed that based on this systemic review, being tackled is the highest rate in phases of play. Lastly, player position did not influence the rate of injury based on this study. Nevertheless, a player with the proper training and good exposure may help the player avoid from getting injuries. Thus, understanding the different type of injury appear in rugby union may beneficial to first aider and prevention system in lesser the injuries rate to occur.

References

1. Garraway, M., Lee, J., Hutton, S., Russell, E., & Macleod, D. (2000). Impact of professionalism on injuries in rugby union. *Journal of Sport Medicine*. https://doi.org/10.1136/bjsm.34.5.348.
2. Kay, E., Kakarla, P., Macleod, D., & McGlashan, T. (1990). Oro-facial and dental injuries in club rugby union players. *Journal of Sport Medicine*. https://doi.org/10.1136/bjsm.24.4.271.
3. Silver, J. (1992). Injuries of the spine sustained during rugby. *Journal of Sport Medicine*. https://doi.org/10.1136/bjsm.26.4.253.

4. Junge, A., Cheung, K., Edward, T., & Dvorak, J. (2004). Injuries in youth amateur soccer and rugby payers comparison of incidence and characteristic. *Journal of Sport Medicine*. https://doi.org/10.1136/bjsm.2002.003020.
5. Scher, A. (1991). Catastrophic rugby injuries of the spinal cord: changing patterns of injury. *Journal of Sport Medicine*. https://doi.org/10.1136/bjsm.25.1.57.
6. Bathgate, A., Best, J., Craig, G., & Jamieson, M. (2001). A prospective study of injuries to elite Australian rugby union player. *Journal of Sport Medicine*. https://doi.org/10.1136/bjsm.2007.037499.
7. McCoy, G. F., Piggot, J., Macafee, A. L., & Adair, I. V. (1984). Injuries of the cervical spine in schoolboy rugby football. *The Bone & Joint Journal, 66*(4), 500–503.

Preferred Coaching Behaviours Among Malaysian KARISMA 2015 Athletes

Tan Chee Hian, Muhammad Aman Bin M. Maamor, Syed Mukhris Bin Syed Adnan and Jong Li Ling

Abstract **Purpose**: This study was to justify the preferred coaching behaviors of Malaysian youth athletes. **Method**: It was a descriptive and inferential research designed using the Leadership Scale of Sports (LSS, Chelladurai [Journal of sport Psychology, 6(1), 27–41, 1994, 2]) questionnaire to conduct and elaborate the related information with self-administration. The samples were purposively selected. The estimated population was about 2500 Malaysian youth athletes in KARISMA 2015 tournament. Referred to Baumgartner and Hensley [Conducting & Reading Research in Kineaiology, McGraw Hill, 2004,1], the appropriate samples size came to 342 samples. The adopted questionnaire ($r = 0.89$) consisted of two sections. Section A: Demographic data of samples which consisted of gender, type of sports, and past experience involvement and achievement in particular sports. Section B was questions perceived that preferred coaching behavior of selected samples. LSS was inclusive dimensions of training and instruction, autocratic, democratic, social support, and positive feedback of coaches' behavior. **Results**: Descriptive statistic showed the highest coaching behavior preferred by overall Malaysian youth athletes was democratic behavior with a mean score of $M = 4.43$, $\mathrm{SD} = 0.29$. Inferentially independent t-Test to compare preferred coaching behavior between gender showed less performed sport female youth preferred autocratic where else, higher performed youths from various sports preferred democratic coaches. ANOVA and Spearman Rho were used to compare the preferred coaching behavior between various sports. It was statistically shown that there was significant difference between sports and coaching behavioral precisely with ranking of overall Malaysian youth athletes preferred democratic coaching in training process. However, it showed that different sequence ranks were preferred between higher performed sports' youth athletes and lower performed sports youth athletes with respect to democratic, social support, positive feedback and on the other hand, autocratic, social support, and positive feedback were preferred. **Conclusion**: Malaysian youth

T. Chee Hian (✉) · M. A. B. M. Maamor · S. M. B. Syed Adnan
Universiti Teknologi MARA, Selangor, Malaysia
e-mail: tanchee@salam.uitm.edu.my

J. Li Ling
Universiti Teknologi MARA, Shah Alam, Sarawak, Malaysia

N. Sulaiman et al. (eds.), *Proceedings of the 3rd International Colloquium on Sports Science, Exercise, Engineering and Technology*,
https://doi.org/10.1007/978-981-10-6772-3_6

athletes preferred training and instruction with democratic of their coaches but lower performed sports youth athletes preferred autocratic rather than democratic comparatively and the rest were sequencing similar. **Contribution**: Results were in line with the past studies and contributed to the coaching science, the body of knowledge.

Keywords KARISMA tournament · Malaysian youth athletes · Training and instruction · Autocratic · Democratic · Social support · Positive feedback · Higher and lower performed sport

1 Introduction

This study is to justify the preferred coaching behaviors of Malaysian youth athletes. This study revisited the preferred coaching behaviors between different gender and different sports. The demographic factors included the genders, type of sport, and the past experience of the Malaysian youth athletes.

Coaches' behaviors are very important factors in a team or individual in order to achieve success because coach's behaviors providing direction, implementing plans, and motivating their athletes especially youth group that could profoundly impact on the team to create a strong and effective team. Coaches behavioral must align the team which, behind the ideas and obtain the group's wholehearted commitment to the overall plan.

Out of the Multidimensional Model of Leadership, Chelladurai and Saleh [3] developed coach's Leadership Scale for Sports (LSS) in which they administered items drawn from scales to physical education and university athletes in various sports. From these items, they then developed five dimensions of coaches' behavioral which was in the dimension of training and instruction being highlighted in this study.

Problems arise with many coaches' behaviors which were talented and skillful in respective sports. Success in the certain sport still depends on both sides which mean the coach and the athlete as well. Coaches' behaviors have an impact on how effectively an organization reaches its objectives.

According to Terry and Howe [4], both the performance and preferred of an athlete can be either enhanced or diminished by the effects of the personality and coaches' behaviors of the teams. Furthermore, athletes come with a very different background of life, social, individual and family values. The exposures toward coaching behaviors are different. Hence, every athlete has a different attitude towards their own coaches' behavior and different perception of increasing on their confidence level. By identifying the preferred coaching behavior, the problem between athletes and coach will be fixed.

This present study divided its objectives into several domains: To identify the coaching behaviors preferred by Malaysian KARISMA athletes; To compare preferred coaching behaviors between different genders; To compare the preferred coaching behaviors between different sports involved Malaysian youth athletes. Hence, it has been hypothesized into two hypotheses: Ho1—there is no significant differences in mean score between preference on coaches' behaviors between gender and Ho2—there is no significant differences between athletes preference coaches behaviors between various sports.

As a clarification, Malaysian youth athletes mean student-athletes in any universities campuses who participated in KARISMA 2015 tournament. Operationally, there were 352 athletes which consisted of 184 of males and 168 females youth athletes. Where else, the coach is any man or woman, paid or unpaid, who carries the responsibility for instructing and directing a sports team.

Coaches' behaviors involved the process of influencing the activities of an organized group toward goal setting and goal achievement. According to Hemphill and Coons [5], leadership as the behavior of an individual when he is directing the activities of a group towards a shared goal. Coach performs two functions. First, the leader performs the relationship function, which includes the feelings, attitudes, and satisfaction of the members of the group and fulfilling an emotional role for the subordinates. Second, leaders are concerned about the task function. The coach must guide the group in the direction of goal attainment, which is one of the primary roles of the influence process [6].

The type of sports athletes participated in this study has also been shown to indicate differences in preferences and perceptions [6]. Examined a variety of sports types and did find that high school athletes preferred more training and instruction compared with midget athletes, and that training and instruction strategies increased slightly from high school to college athletes. Terry and Howe [4] even examined differences between athlete preferences based on their participation in either interdependent or independent sports. The study found that independent athletes preferred democratic behaviors in coaches, while non-independent athletes preferred autocratic behavior. Using the Leadership Scale of Sport (LSS), studied found that team sport athletes preferred training and instruction behavior significantly more than members of individual sports [3].

2 Method

This study used a descriptive and inferential research by using the questionnaire to conduct this study to find the related information. This study directly looked at the process of answering the questionnaire, and this is because the samples have 10–15 min to answer the questionnaires. Krejcie and Morgan [7] said appropriate samples come to 342 ($n = 342$). The questionnaire was distributed with the total of 352 Malaysian youth athletes from various sports in case of dropped out rate concerned.

3 Samplings

The sample was selected purposively. Target estimated population was 2500 of KARISMA Malaysian youth athletes. The total questionnaire was distributed to 184 samples of male athletes and 168 of female athletes. Data were collected in 6 weeks.

4 Instrumentation

This study conducted was adopted an established questionnaires. The questionnaire consists of two sections. Section A is about demographic section of samples. Section B is on coaching behaviors. It was adopted Leadership Scale of Sports (LSS) and it included the training and instruction with autocratic behavior, democratic behavior, social support, and positive feedback. Samples indicated on a 5-point Likert with the reliability of 0.89.

5 Data Analysis

The collected data had been analyzed by using descriptive statistic and an inferential statistic which are Independent *t*-Test, ANOVA, and Spearman. Descriptive data determine the overall mean score of coaching behaviors preferred by Malaysian youth athletes.

Moreover, an inferential statistic which is Independent *t*-Test for the comparison between genders and ANOVA for the comparison between sport and Spearmen Rho in order to rank the preferred coaching behaviors of all samples.

6 Results

The overall ranking of coaches behaviors preferred by Malaysian KARISMA Athletes as showed in the Table 1.

Table 1 showed the overall level of preferred coaching behavior by KARISMA 2015 athletes based on mean and standard deviation between dimensions of coaching behaviors liked: Positive Feedback, Social Support, Democratic, Autocratic, and Training Instruction. Democratic with the highest score of $M = 4.43$, and the lowest is social support with the score $M = 3.44$. Based on this data, the highest coaching behavior preferred by 2015 Malaysian KARISMA athletes' was Democratic.

Table 1 Descriptive result of preferred coaching behaviors of Malaysian KARISMA athletes ($n = 342$)

Coaching behaviors	Minimum	Maximum	Mean	Ranking
Democratic	3	5	4.43	1
Positive feedback	3	5	4.31	2
Training instruction	3	5	4.23	3
Autocratic	2	5	3.90	4
Social support	2	5	3.44	5

Table 2 Independent t-Test result of preferred coaching behavior between gender

Coaching behaviors	Sex	N	t	Sig. (2-tailed)
Autocratic	Male	184	2.21	0.027
	Female	168	2.23	0.026
Social support	Male	184	0.61	0.541
	Female	168	0.61	0.541
Training instruction	Male	184	0.39	0.698
	Female	168	0.39	0.696
Positive feedback	Male	184	0.31	0.756
	Female	168	0.31	0.756
Democratic	Male	184	0.11	0.915
	Female	168	0.11	0.914

$P = < 0.05$ show significant

Table 2 showed the result of preferred coaching behavioral based on preferred coaches behaviors. The statistical reading for male and female youth athletes was significantly different between Autocratic behavior among less performed youth athletes.

Comparison of Coaching Behavioral preferred by 2015 Malaysian KRISMA athletes between sports (Table 3).

Table (4) showed the ANOVA result of preferred coaching behavior for Training Instruction ion between different sports. The statistical reading is $F = 6.40$, $P < 0.05$.

7 Discussion

Based on Table 3, the result concludes that the highest preference of coaching behavior overall Malaysian youth athletes was democratic. Meanwhile, the lowest preference was social support behavior. It reflected that the democratic behavior was the most effective coaching behavior preferred by the youth athletes in line with [8] studied. However, democratic coaches do not have to leave all the decision-making up to the team.

Table 3 Preferred coaching behaviors between different sports

Type of sport		Training instruction	Autocratic	Democratic	Social support	Positive feedback
Football	Mean	4.22	3.10	4.46	3.31	4.56
	Standard deviation	0.31	0.32	0.28	0.29	0.40
Netball	Mean	4.23	3.93	4.49	3.41	4.01
	Standard deviation	0.31	0.33	0.24	0.26	0.31
Handball	Mean	4.24	4.14	4.51	3.30	4.23
	Standard deviation	0.22	0.32	0.28	0.34	0.44
Volleyball	Mean	4.19	3.88	4.35	3.26	4.31
	Standard deviation	0.40	0.46	0.31	0.33	0.53
Hockey	Mean	4.23	3.93	4.46	3.09	4.33
	Standard deviation	0.20	0.36	0.27	0.47	0.29
Rugby	Mean	3.90	3.66	4.53	3.39	4.28
	Standard deviation	0.37	0.44	0.20	0.38	0.25
Basketball	Mean	4.22	3.69	4.32	3.49	4.48
	Standard deviation	0.28	0.28	0.17	0.34	0.51
Futsal	Mean	3.98	3.76	4.34	3.37	4.25
	Standard deviation	0.25	0.33	0.37	0.42	0.50
Silat seni	Mean	4.13	3.88	4.35	3.45	4.33
	Standard deviation	0.20	0.28	0.20	0.25	0.47
Taekwando	Mean	3.95	3.59	4.40	3.45	3.94
	Standard deviation	0.33	0.46	0.17	0.28	0.32
Petanque	Mean	4.02	3.80	4.21	3.09	4.60
	Standard deviation	0.25	0.38	0.34	0.38	0.43

The finding of this study was showed the result of the second objective which was the preferred coaching behavior among Malaysian youth athletes between different sports.

Table 4 ANOVA result of preferred coaching leadership styles between sports

Coaching behaviors		*F*	Df	Sig.
Autocratic	Between groups	10.09	21	0.000
	Within groups		330	
Training instruction	Between groups	6.40	21	0.000
	Within groups		330	
Social support	Between groups	5.48	21	0.000
	Within groups		330	
Positive feedback	Between groups	4.74	21	0.000
	Within groups		330	
Democratic	Between groups	1.59	21	0.050
	Within groups		330	

$P = < 0.05$ show significant

8 Conclusion

This study concluded that the preferred coaching behavior of Malaysian youth athletes was democratic. Hence, coaches needed to implement the democratic coaching behavior towards their athletes and maybe this method will increase athletes' performance in the future. This study is also in line with some past studies and it could contribute to coaching science, the body of knowledge.

References

1. Baumgartner, T. A., & Hensley, L. D. (2013). *Conducting & reading research in kineaiology* (5th ed). McGraw Hill.
2. Chelladurai, P. (1994). Discrepancy between preferences and perceptions of leadership behavior and satisfaction of athletes in varying sports. *Journal of Sport Psychology, 6*(1), 27–41.
3. Chelladurai, P., & Saleh, S. (1980). Dimensions of leader behavior in sports: Development of a leadership scale. *Journal of Sport Psychology, 2*(1), 34–45.
4. Terry, P., & Howe, B. (1984). Coaching preferences of athletes. *Canadian Journal of Applied Sport Sciences, 9*(4), 188–193.
5. Hemphill, J. K., & Coons, A. E. (1957). Development of the leader behavior description questionnaire. *Leader behavior: Its description and measurement, 6,* 38.
6. Turman, P. D. (2000). *It's all in the timing: The examination of athletes' preferences and perceptions of their coaches' leadership strategies throughout the course of an athletic season.* (9973605 Ph.D.), The University of Nebraska—Lincoln, Ann Arbor. Retrieved from http://search.proquest.com.ezaccess.library.uitm.edu.my/docview/304612654?accountid=42518ProQuestDissertations&ThesesGlobaldatabase.
7. Krejcie, R. V., & Morgan, D. W. (1970). Determining sample size for research activities. *Educational and psychological measurement, 30*(3), 607–610.

8. Pratt, S. R., & Eitzen, D. S. (1989). Differences in coaching philosophies between male coaches of male and female basketball teams. *International Review for the Sociology of Sport, 24*(2), 151–161.

Calisthenics and Passive Stretching Exercises for Patients with Type 2 Diabetes Mellitus: A Study Protocol

Norazila Nordin and Zainal Abidin Zainuddin

Abstract Resistance training using one's own body weight has a pivotal role in lessen the risk factors associated with type 2 diabetes mellitus (T2DM). One of the typical resistance training exercises is traditional calisthenics, frequently included in rehabilitation or physical therapy program. Likewise, passive stretching is beneficial in improving glycemic control and physical fitness. To date, however, the effect of calisthenics exercises and passive stretching that could improve the risk factors in the population are not well understood. The propose study therefore, sets out to assess the effectiveness of the supervised calisthenics exercise and passive stretching on glycemic controls, body composition, risk factors that associate with cardiovascular disease, and some fitness parameters. Adult patients who are registered in the Health Centre, UTM, Skudai and meet some inclusion criterion will be randomly chosen for this study. A total of 66 type 2 diabetics will be randomised into three experimental groups (i.e. calisthenics (CS) and passive stretching (PS) and also combined exercise (CP) and will receive treatment for 12 weeks of study duration. All parameters will be measured at baseline and post measurement. We are aiming for an improvement in physical measurement and blood biomarkers in subjects, besides investigating the effectiveness of intervention exercises on fitness parameters thereby improving their quality of life.

Keywords Calisthenics exercises · Passive stretching · Type 2 diabetes mellitus

1 Background of the Study

Diabetes Mellitus (DM) is one of predominant cardiovascular risk factor [1]. The prevalence of diabetes mellitus has been estimated for each country for the years 2010 and 2030, thereby, it was manifested that Malaysia was the highest prevalence

N. Nordin (✉) · Z. A. Zainuddin
School of Education, Faculty of Social Sciences and Humanities, Universiti Teknologi Malaysia, Skudai, Malaysia
e-mail: azielanordin79@gmail.com

N. Sulaiman et al. (eds.), *Proceedings of the 3rd International Colloquium on Sports Science, Exercise, Engineering and Technology*,
https://doi.org/10.1007/978-981-10-6772-3_7

of T2DM among Asian countries [2]. At that time, the comparative prevalence had shown that Malaysia was under top ten among countries that affecting about 11.6% in 2010 and it is predicted to increase to 13.8% in 2030 [2]. In the national level, prevalence of DM is projected to increase among the adult aged 30 years and older, from 8.3% in 1996 to 14.9% in 2006 [3] and 20.8% in 2011 [4] have seen in Malaysia. This is as a result of changing population of demographics (e.g. ageing and urbanization), lifestyle (dietary pattern and physical inactivity) and also increase in obesity [5]. As a matter of fact, T2DM is common disorder in the older person as prevalence increases with age [1].

In order to prevent some kind of hypokinetic diseases and maintaining physical fitness, adult particularly aged range 18–64, should engage physical activity includes leisure time physical activity, working, household chores, or planned exercises [6]. The exercise intervention strategies, designed for the population of type 2 diabetics should be adopted for each person and based on comorbidities, contraindication and individual goals [7]. Most scientific organization recommended that a minimum of 150 min of moderate to vigorous intensity of aerobic activity, at least 3 days a week. Resistance training (RT) exercises are recommended for at least 2 days a week. On the other hand, flexibility exercise may be part of the intervention program as complement to the other types of exercise [8].

RT has beneficial effects in enhancing physical fitness including musculoskeletal strength, endurance power and conditioning of an individual. More over, it is encompasses a wide range of modalities including free weight, weight-stacked machine, medicine ball, resistive band and body weight (calisthenics) [9–11]. The present study defines calisthenics as exercise that apply isotonic and isometric muscle exercises that target to all major muscles groups, in order to gain muscle fitness, flexibility and improve metabolic profiles.

Calisthenics exercise directly targets the underlying pathophysiology of metabolic syndrome, visceral obesity, improving lipid profiles and cardiac indices in adult population [12–16]. The unique constellation of benefits of this training regimen for instance could improve muscle strength, flexibility, proprioception, coordination and balance, more congruent and effective to any other exercise regimen and diet modification strategy [17–19].

Calisthenics exercise offers a radically different non-pharmachologic approach to the treatment of T2DM and its risk factors. As oppose to aerobics exercise, notably with low to moderate intensity, this exercise could improve metabolic risk factors particularly in obese, type 2 diabetics. Result of previous study has proved that with low-intensity RT, improved glycemic controls, body fat mass, body fat percentage and HDL-c [20].

Flexibility exercises are recommended to complement the intervention exercise and can benefit older people with type 2 diabetes [7, 8]. Nelson and Kokkonen [21] added another adjunct effect of regular stretching exercise that could improve the level of blood glucose. It was manifested by the earlier study revealed that passive stretching exercises have improved blood glucose level in adult with Type 2 Diabetes [22–25]. The movement of passive stretching is imposed by external force [22, 23] without contribution to generate force [26]. This study, however, defines passive stretching as the range of motion imposed by someone force to assist stretch with no effort by the subject being stretched.

The beneficial effects of RT on lowering cardiovascular risk (i.e. blood pressure and lipids) have been reviewed elsewhere [27, 28]. Of the studies reviewed here, the impact of RT (especially the usage of some equipment such as weight-stacked machine, free weight and resistive bands) manifested some improvements on metabolic risk factors in T2DM [20, 29–33]. Despite of high-quality equipment with longer intervention duration indicated a significant improvement in muscle strength, however, there were no significant improvement been observed in some glycemic controls (HbA_{1c}) and lipid profiles (HDL-c) [34–36] and BMI [37].

Despite the previous publications have recognized calisthenics as supervised physical exercise, there is a notable paucity of evidence-based literature describing the impact of calisthenics exercise and passive stretching on risk factors associated with type 2 diabetes. In an attempt to shed some light on these issues, the present study would be concentrating on calisthenics exercise, passive stretching and combined exercise on glycemic controls, body composition, lipid profiles, cardiac indices, flexibility level and muscle strength in patients with Type 2 Diabetes Mellitus.

2 Methodology

2.1 *Design and Recruitment*

We designed a quasi-experimental study among patients with T2DM. All subjects will be recruited using purposive sampling, a non-probability sampling method. All potential patients will be traced and identified based on the eligibility criterion. The inclusion and exclusion criteria are depicted in Table 1. Those patients with T2DM at the general practices fulfilling the inclusion criteria by the physician, will be invited for screening for their eligibility. The recruitment further involves a health and physical examination, which will be conducted by medical physicians. The physicians will assess all medical conditions including cognitive function, cardiovascular condition and history of injury (if any) and other medical conditions. The potential candidates will be recruited, after the confirmation from the physicians about their safety that will not be compromised.

2.2 *Sample Size Estimation*

It was calculated that 66 patients with T2DM would be recruited this study to demonstrate a significant difference at 95% power and 5% significance. The power calculation were based on the literature [36] and aimed at a difference of 0.5% in HbA_{1c}

Table 1 Eligibility criterion for T2DM

Inclusion	Exclusion
1. Registered patient (regular)	1. Significant cognitive impairment
2. Under treatment; Diagnosed with T2DM	2. Non-ambulatory
3. Clinically stable T2D (no change in medication for the previous 2 months)	3. Lower extremity amputation (other than toes)
4. Voluntarily/willing to come to sports science lab	4. Alcohol and substance abuse
5. HbA1c range of $\geq$6.5%	5. Inability to comply the present study requirement
6. Abnormal lipid profiles	6. Unstable CVD diseases
7. Able to perform all movements and assessment	7. Unrepaired aortic aneurysm
8. Understand the study purposes	8. Proliferative diabetic retinopathy
9. Sedentary (no structured exercises: < 1/week)	9. Rapidly progressive terminal illness [40]
	10. Conceived

value in the treated groups. Assuming an alpha level of 0.05 and statistical design of F test of repeated measures (between and within effects). Based on 15% attrition rate for 12 weeks study duration, thus, the sample size was increased to 76.

2.3 *Ethics and Data Security*

This study is comply to the principle of Declaration of Helsinki [38]. The present study was approved by the Medical Research & Ethical Committee (MREC), Ministry of Health (NMRR-16-1505-30985-IRR). All patients will be asked for written consent informed consent according to the standard form.

2.4 *Exercise Protocol*

The exercise protocol is currently based on recommendations for management of T2DM from The American College of Sports Medicine (ACSM) [10] and the American Diabetes Association (ADA) [39] with some current evidence statements, frequency, intensity, duration, mode, and rate of progression [8].

ACSM and ADA recommended 8–10 exercises, with 10–15 repetitions per set. This study, however, adopts 13–14 exercises for calisthenics, with 8–12 repetitions, depending on circumstances. Each subject will be guided and supervised to perform the exercises for 3 times per week at the sports science laboratory, and another 2 additional times will be at their respective home. The subjects are advised to perform the exercises on their own at their respective houses when they are able to do all the movements correctly and independently. The intervention program will be held for about 12 weeks duration time.

Both calisthenics and passive stretching exercises target the large muscle groups of upper limb, lower limb, and the trunk. For calisthenics exercises for instance, consists of *side lying abduction, lunges, high lunges, plank knee pull in, squat, push-ups, plank rotation, dips, hand walk-out, double leg abs press, v-crunch, quadruped, plank, and side plank*. It will take approximately 40 min, per se. A complete exercises manual (illustrate) will be provided to each subject. For combined group for instance, consists of 6 phases (warm-up, lower body, upper body, trunk, passive stretching).

Speaking at the exercise movement, it begins with slow to moderate, constant pace. The rest between sets is not to be longer that 60 s. The intensity is determined by relative intensity of subject self-rating, adopting the Borg Relative Perceived Exertion (RPE) Scale from 6 to 20. In order to reach the moderate intensity level, the score of each subject has to be within 11–13.

To monitor subject's compliance, a log book will be given to each of them as a compliance monitoring approach. It will be used to record any sign or symptom, resting blood pressure, resting heart rate, and the exercise heart rate.

In addition to the passive stretching, all movements (e.g. *Supine knee flexor-plantar flexor, Prone hip flexor, Recumbent knee, ankle and back stretch, Reclining lower-trunk extensor, Prone lower-trunk flexor, Seated hip adductor stretch, Seated shoulder lateral flexor, Seated hip external rotators, extensors*) will be guided by the experimenter or the partner. All stretches will be done passively with the help of the partner or experimenter. Each stretch will be held for about 30 s, then rest for 15 s between stretching. The intensity level on the scale is from 1 to 3, with light pain sensation [21, 22].

2.5 *Outcome Measures*

The primary outcome measure are glycated hemoglobin (HbA_{1c}) and fasting blood glucose (FBG). Secondary outcomes are fat.%, Muscle mass, BMI, fat free mass, systolic blood pressure (SBP), diastolic blood pressure (DBP), resting heart rate (RHR), low density lipoprotein (LDL-c), high density lipoprotein (HDL-c), triglycheride (TG), total cholesterol (TC), muscle strength level and flexibility level. All outcome measures are collected at baseline and post intervention (12 weeks).

3 Discussion

Exercise intervention was seen as a cornerstone for non-pharmachological treatment, particularly of type 2 Diabetes, recognized by most international scientific organizations: International Diabetes Federation (IDF), American Diabetes Association (ADA) and American College of Sports Medicine (ACSM) [7]. The development of the intervention design needs through the process of planning the structure and the implementation of exercise chosen are based on the training principles for instance frequency, intensity, time (duration) and progression [10]. More over, some variables need to be considered in the development of exercise prescription for T2DM patients including safety factors (clinical status, exercise capacity), associated factors (musculoskeletal limitation) and personal health and fitness goal [10]. To gain feasibility of the designed training program, the pilot study will be conducted beforehand.

4 Conclusion

This study would make a unique contribution to public health in the working population. In addition, this study will contribute to improve the quality health care for diabetics in health clinic and would recommend a multifactorial approaches emphasizing education on patient behavior, glycemic controls, body composition, risk factors status, individualize physical fitness, and improved personal behavior towards physical exercise.

Acknowledgements This work was supported by the health practitioners of the UTM Health Clinic, with the cooperation of the Rehabilitation Centre, Sultan Ismail Hospital Johore Bahru, Malaysia.

References

1. Holt, R. I. G., & Hanley, N. A. (2012). *Endocrinology and Diabetes* (6th ed.). Oxford. UK: Wiley Blackwell.
2. International Diabetes Federation. (2009). *IDF Diabetes Atlas*, 4th Edn. Retrieved from http://www.idf.org/diabetesatlas/downloads.
3. Letchuman, G. R., Wan Nazaimoon, W. M., Wan Mohamad, W. B., Chandran, L. R., Tee, G. H., Jamaiyah, H., et al. (2006). Prevalence of Diabetes in the Malaysian National Health Mobidity Survey III. *Medical Journal Malaysia, 65*(3).
4. National Health and Morbidity Survey. (2011). *The Fact Sheet*. Retrieved from www.moh.gov.my.
5. Holt, R. I. G., & Hanley, N. A. (2012). *Endocrinology and Diabetes* (6th ed.). Wiley Blackwell: Oxford. UK.
6. WHO. (2011). Global Recommendation on Physical Activity For Health. Retrieved from http://www.ncbi.nlm.nih.gov/books/NBK305057/.
7. Mendes, R., Sousa, N., Almeida, A., Subtil, P., & Guedes-Marques, Fdo. (2015). Victor Machado Reis, José Luís Themudo-Barata. Exercise prescription for patients with type 2 diabetes—a synthesis of international recommendations: Narrative review. *British Journal Sports Medicine, 1,* 1–4.
8. Colberg, S. R., Sigal, R. J., Fernhall, B., Regensteiner, J. G., Blissmer, J. G., Rubin, R. R., et al. (2010). Exercise and Type 2 Diabetes. The American College of Sports Medicine and American Diabetes Association: Joint Position Statement. *Diabetes Care, 33*(12), e147–e167.
9. Howley, E. T., & Thompson, D. L. (2012). *Fitness Professional's Handbook* (6th ed.). USA: Human Kinetics.

10. American College Of Sports Medicine. (2010). *ACSM's guidelines for exercise testing and prescription* (8th ed.). Baltimore: William and Wilkins.
11. Fleck, S. J., & Kraemer, W. J. (1997). *Designing resistance training program* (2nd ed.). Champaign, IL, US: Human Kinetics.
12. Tsuzuku, S., Kajioka, T., Endo, H., Abbott, Robert D., Curb, J. D., & Yano, K. (2007). Favorable effects of non-instrumental resistance training on fat distribution and metabolic profiles in healthy elderly people. *Original Article European Journal of Applied Physiology, 99,* 549–555.
13. Guzel, N. A., Pinar, L., Colakoglu, F., Karacan, S. & Ozer, C. (2012). Long-Term Callisthenics Exercise Related Changes in Blood Lipids, Homocysteine, Nitric Oxide Level, and Body Composition in Middle-aged Healthy Sedentary Women. *Chinese Journal of Physiology.* 55.xx.
14. Ajayi-Vincent, O. B., & Adesina, M. O. (2013). Effects of resistance training on the blood lipid variables of young adults. *European Scientific Journal, 9* (12).
15. Teramoto, M., & Golding, L. A. (2013). Regular Exercise and Plasma Lipid Levels Associated With the Risk of Coronary Heart Disease. *Research Quarterly For Exercise and Sports., 80*(2), 138–145.
16. Sarvan, S. (2013). The Effect eight-week Core Exercises on Blood Lipid Profiles In Females. *Australian Journal of Basic and Applied Sciences, 7*(10), 209–214.
17. Ayan Perez, C., Martin Sanchez, V., Teixeira, De Souza, & De Paz Fernandez, J. A. (2007). Effects of a Resistance Training Program in Multiple Sclerosis Spanish Patients: A Pilot Study. *Journal of Sports Rehabilitation, 16,* 143–153.
18. Ozer Kaya, D., Duzgun, I., Baltaci, G., Karacan, S., & Colakoglu, F. (2012). Effects of Calisthenics and Pilates Exercise on Coordination and Proprioception in Adult Women: Randomized Controlled Trial. *Journal of Sports Rehabilitation., 21,* 235–243.
19. Farinatti, P. T. V., Rubini, E. C., Silva, E. B., & Vanfraechem, J. H. (2014). Flexibility of the Elderly After One-Year Practice of Yoga and Calisthenics. *International Journal of Yoga Therapy, 24,* 71–77.
20. Hamasaki, H., Kawashima, Y., Tamada, Y., Furuta, M., Katsuyama, H., Sako, A., et al. (2015). Associations of Low-intensity resistance training with body composition and lipid profiles in obese patients with Type 2 Diabetes. *PLoS ONE, 10*(7), e0132959.
21. Nelson, A. G., & Kokkonen, J. (2014). *Stretching Anatomy* (2nd ed.). United States: Human Kinetics.
22. Nelson, A. G., Kokkonen, J. & Arnall, D. A. (2011). Twenty minutes of passive stretching lowers glucose levels in an at-risk population: An experimental study. *Journal of Physiotherapy, 57.*
23. Putri Thanaya, S. A., Indrayani, A. W., & Andayani, N. L. N. (2015). Passive Stretching Menurunkan Kadar Glukosa Darah Pada Penderita Diabetes Melitus Tipe 2 Di Kota Denpasar. *Majalah Ilmiah Fisioterapi Indonesia, 3*(1).
24. Park, S. H. (2015). Effects of Passive Static Stretching on Blood Glucose Levels in Patients with Type 2 Diabetes Mellitus. *Journal of Physical Therapy Science., 27,* 1463–1465.
25. Solomen, S., Syakya, R., Agarwal, K., Aaron, P., & Pradeep, S. (2015). Passive Stretching Versus Active Stretching on Immediate Blood Glucose in Subjects With Type II Diabetes Mellitus—A Pilot Study. *International Journal of Physical Education, Sports and Health., 2*(1), 146–149.
26. Alter, M. J. (1996). *Science of Flexibility* (2nd ed.). Champaign, IL, US: Human Kinetics.
27. Hayashino, Y., Jackson, J. L., Fukumori, N., Nakamura, F., & Fukuhara, S. (2012). Effects of Supervised Exercise on Lipid Profiles and Blood Pressure Control in People with Type 2 Diabetes Mellitus: A Meta-analysis of Randomized Controlled Trials. *Diabetes Research and Clinical Practice, 98,* 349–360.
28. Gordon, B. A., Benson, A. C., Bird, S. R., & Fraser, S. F. (2009). Resistance Training Improves Metabolic Health In Type 2 Diabetes Mellitus: A Systemic Review. *Diabetes Research and Clinical Practice, 83,* 157–175.
29. Honkola, A., Forsen, T., & Ericksson, J. (1997). Resistance Training improves the metabolic profile in individuals with type 2 diabetes. *Acta Diabetologica, 34,* 245–248.
30. Castaneda, C., Layne, J. E., Munoz-Orians, L., Gordon, P. L., Walsmith, J., Foldvari, M., et al. (2002). A Randomized Controlled Trial of Resistance Exercise Training to Improve Glycemic Control in Older Adults With Type 2 Diabetes. *Diabetes Care, 25*(12), 2335–2341.

31. Bweir, S., Al-Jarrah, M., Almalty, A. M., Maayah, M., Smirnova, I. V., Novikova, L., et al. (2009). Resistance Exercise Training Lowers HbA1c More Than Aerobic Training in Adults with Type 2 Diabetes. *Diabetology & Metabolic Syndrome., 1,* 27.
32. Ghalavand, A., Shakeriyan, S., Monazamnezhad, A., & Delaramnasab, M. (2014). The Effect of Resistance Training on Cardio-Metabolic factors in Males with Type 2 Diabetes Jundishapur. *Journal Chronic Disease Care, 3*(4), e23346.
33. Dadgostar, H., Firouzinezhad, S., Ansari, M., Shima Younespour, S., Azam Mahmoudpour, A., & Khamseh, M. E. (2016). Supervised group-exercise theraphy versus home-based exercise therapy: Their effects on quality of life and cardiovascular risk factors in women with type 2 diabetes. *Diabetes & Metabolic Syndrome: Clinical Research & Reviews., 2016,* 1871–4021.
34. Plotnikoff, R. C., Eves, N., Jung, M., Sigal, R. J., Padwal, R., & Karunamuni, N. (2010). Multicomponent, Home-based Resistance Training for Obese Adults with Type 2 Diabetes: A Randomized Controlled Trial. *International Journal of Obesity, 34,* 1733–1741.
35. Hazley, L., Ingle, L., Tsakirides, C., Carroll, S., & Nagi, D. K. (2010). Impact of a short-term, moderate intensity, lower volume circuit resistance training programme on metabolic risk factors in overweight/obese Type 2 Diabetics. *Research In Sports Medicine., 18,* 251–262.
36. Arimi Fitri, M. L. (2013). *Effects of a 16-week, home-based, high-intensity, progressive-resistance training program on glucose homeostasis, cardiovascular disease risk factors, and functional capacity of older Malaysians with type 2 Diabetes Mellitus. Doctor Philosophy.* Serdang: Universiti Putra Malaysia.
37. Ibaneze, J., Izquierdo, M., Arguelles, I., Larrion, J. L., Garcia-Unciti, M., Idoate, M., et al. (2005). Twice-weekly Progressive Resistance Training (PRT) Decreases Abdominal Fat & Improves Insulin Sensitivity in Older Men With Type 2 Diabetes. *Diabetes Care, 28*(3), 662–667.
38. Declaration of Helsinki. (2008). *World Medical Association.* Retrieved on 27 March 2016 from: http://www.who.int/bulletin/volumes/86/8/08-050955/en/.
39. American Diabetes Association. (2014). Standard of Medical Care in Diabetes. *Diabetes Care, 37*(1), S14–S80.
40. Simpson, K. A., Mavros, Y., Kay, S., Meiklejohn, J., Vos, N. D., Wang, Y., Guo, Q., et al. (2015). Graded Resistance Exercise And Type 2 Diabetes in Older adults (The GREAT2DO study): methods and baseline cohort characteristics of a randomized controlled trial. *Simpson et al. Trials.*

Level of Stress Between Obese and Nonobese Malaysians

Vincent Parnabas, Julinamary Parnabas and Antoinette Mary Parnabas

Abstract Obesity is the excess of body fat, often resulting in impaired health. For many years, people have suspected that stress and obesity are linked but since lack of research, it is not easy to determine the relationship. The aim of this research was to identify the level of stress between obese and nonobese Malaysians. The sample of this study consisted of 121, obese ($N = 65$) and nonobese ($N = 56$) Malaysians residing Shah Alam, Selangor. The sample was differentiated between obese and nonobese based on their Body Mass Index (BMI) result. BMI of 30 kg or more are classified as obese. The instrument used in this study comprised of a 73-item Stress Indicator Questionnaire (SIQ). The result showed that the level of stress on obese higher than nonobese, $t\ (121) = 13.3120$, $p < .01$. Besides that, all the symptoms of stress including Physical ($\bar{x} = 20.7343$), Sleep ($\bar{x} = 19.2987$), Behavioural ($\bar{x} = 21.5171$), Emotional ($\bar{x} = 23.3312$), and Personal Habits ($\bar{x} = 15.4591$) were higher on obese compared to nonobese respondents. Even though the present study showed that food most probably perceived as a source to reduce stress but obese population should use other strategies to reduce their stress level than depending solely on food.

Keywords Body mass index (BMI) · Obese · Nonobese · Stress

V. Parnabas (✉)
Sport Science and Recreation Faculty, University of MARA Technology (UiTM), Shah Alam, Malaysia
e-mail: vincent@salam.uitm.edu.my; vincentbarnabas@yahoo.com

J. Parnabas
Institut Pendidikan Guru, Kampus Darulaman, Jitra, Kedah, Malaysia

A. Mary Parnabas
Medical Unit, Hospital Taiping, Taiping, Perak, Malaysia

N. Sulaiman et al. (eds.), *Proceedings of the 3rd International Colloquium on Sports Science, Exercise, Engineering and Technology*,
https://doi.org/10.1007/978-981-10-6772-3_8

1 Introduction

Several decades in the past media has focused public attention on the food consumed and its influence on physical and mental well-being [3]. Generally, it is assumed that obesity result from overeating. Obesity is defined as unhealthy amount of body fat [10]. Obesity is an abnormal accumulation of body fat, usually 20% or more over an individual's ideal body weight [13]. Obesity is the excess of body fat, often resulting in impaired health. According to Obesity Education Initial Task Force of the National Institutes of Health and National Heart, Lung and Blood Institutes [14], overweight and obesity are classified using the body mass index (BMI = weight kg/height square).

Figure 1 showed that individuals with BMI between 25 and 29.9 kg are classified as overweight, and those with a BMI of 30 kg or more are classified as obese.

Obesity contributes to the development of multiple chronic diseases. In fact, the severity of chronic is worst in obese patients. According to the National Heart, Lung and Blood Institute [14], Gilbert [6] and Kolata [11], obese individuals have a higher risk of cardiovascular disease, dyslipidemia, hypertension, lipid disorder, coronary

Classification	BMI(kg/m^2)	
	Principal cut-off points	Additional cut-off points
Underweight	<18.50	<18.50
Severe thinness	<16.00	<16.00
Moderate thinness	16.00 - 16.99	16.00 - 16.99
Mild thinness	17.00 - 18.49	17.00 - 18.49
Normal range	18.50 - 24.99	18.50 - 22.99
		23.00 - 24.99
Overweight	≥25.00	≥25.00
Pre-obese	25.00 - 29.99	25.00 - 27.49
		27.50 - 29.99
Obese	≥30.00	≥30.00
Obese class I	30.00 - 34.99	30.00 - 32.49
		32.50 - 34.99
Obese class II	35.00 - 39.99	35.00 - 37.49
		37.50 - 39.99
Obese class III	≥40.00	≥40.00

Source: Adapted from WHO, 1995, WHO, 2000 and WHO 2004.

Fig. 1 Classification of BMI

heart disease, sleep apnea, glucose intolerance, insulin resistance, gallbladder disease, osteoarthritis, diabetes, high blood pressure, heart disease, and certain types of cancer.

For many years, countless people have suspected that stress and obesity are linked but since lack of research, it is not easy to determine the relationship. Hussien and Hussien [9] defined stress as the situation by which an individual suffers from physical and psychological hypertension resulted from factors that cannot be handled and exceeds human capability to cope. According to Bonnie Taub-Dix in Thompson [17] stress may lead to inappropriate eating habit, which can cause obesity. Furthermore, research of Han et al. [8] on a group of nurses showed that nurses who were obese perceived their career as more stressful compared to nonobese nurses.

According to Van-Strien and Ouwens [18] and Leon et al. [12], stress can make a person to lose awareness on internal physiological states, which can lead the person to not only be able to differentiate between hunger cues and emotional arousal. There are people who eat more when they are in stress or in an emotional situation. Besides that, people who are identified as "emotional eaters" are more vulnerable to gain weight compared to nonemotional eaters [4, 18].

According to General Adaptation Syndrome, there are three stages, a person will go through when experience stress, namely alarm, resistance and exhaustion stage [1]. In the alarm stage, the person starts to identify the existence of a stressful situation and represents the body's initial reaction to stress. During this stage, the body is simulated for the flight or fight responses and this energizes the body to assemble its resources to deal with the threat resulting in physiological arousal. If the stress continues, a person moves from the alarm stage into the resistance stage. The resistance stage is characterized by the body's reaction to continued stress, when it adopts strategies to deal with it. It is a stage when arousal is very high and where the person experiencing stress responds to the physiological and psychological aspects of the stress situations. The reaction to the stressful situation might be in the form of increased worry and anxiety (cognitive state anxiety) or increased physiological reaction, such as increased body tension, nervousness, sweating, and heartbeat (somatic state anxiety). At these stages, for some individuals over eating reduce their stress and anxiety levels. In other words, overeating may be used as coping strategies to fight with the stress situation.

Figure 2 shows a comparison of obese population among Asian countries, highlighted that Malaysia has the highest number of obese compared to Thailand, Singapore, Taiwan, Japan, Philippines, Indonesia, and Vietnam. In Malaysia, overweight and obesity prevalence were 41.1 and 14.4% [19].

The major question that has been asked in research on stress and eating is whether stress increases, decreases or produces no changes in eating behaviours. Therefore, the aim of this research was to identify the level of stress between obese and nonobese Malaysian. In other words, this study aims to find out whether stress can be a factor for overweight among Malaysian.

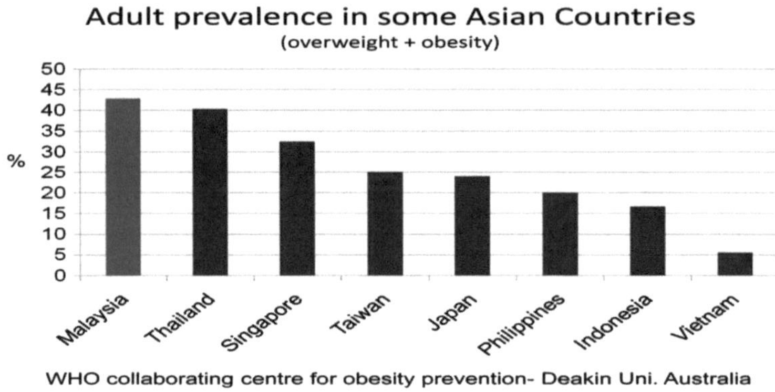

Fig. 2 Comparison of overweight percentage among Asian Countries

2 Method

The sample of this study consisted of 121 respondents, with obese ($N = 65$) and nonobese ($N = 56$). The sample was recruited from the district of Shah Alam, Selangor, Malaysia. The samples were differentiated between obese and nonobese based on their BMI result as below:

BMI Categories:
Underweight = <18.5
Normal weight = 18.5−24.9
Overweight = 25−29.9
Obesity = BMI of 30 or greater.

The instrument used for this study comprised of a 73-item Stress Indicator Questionnaire (SIQ). The questionnaire measured five kinds of stress symptoms, which include physical (21 item), sleep (5), behavioural (17 items), emotional (21 items), and personal habit (9 items).

3 Result

3.1 Profile of the Respondents

Frequency, percentage, mean, and standard deviation are presented in Table 1, which shows the overall results of the obese and nonobese respondents' profile. The profile of the respondents described the ethnics, gender, and age. There were 65 obese and 56 nonobese respondents participated in this study. There were 41 male and 80 female respondents. Based on the ethnic, 51 Malays, 38 Indians, and 32 Chinese respondents.

Table 1 Profile of the obese and nonobese respondents

Variables	Frequency	Percentage	Mean	SD
Ethnics				
Malay	51	42.15		
Indian	38	31.40		
Chinese	32	26.45		
Gender				
Male	41	33.88		
Female	80	66.12		
BMI Categories				
Normal weight	56	46.28		
Obesity	65	53.72		
Age				
Male			57.11	2.52
Female			59.45	2.47
Overall			60.16	2.31

Table 2 Cronbach reliability coefficients

Factors	Cronbach's Alpha ($n = 121$)
Physical indicator	0.8612
Sleep indicator	0.8713
Behavioural indicator	0.8799
Emotional indicator	0.8811
Personal indicator	0.8717

The mean age for overall respondents was 60.16 years old. The age of male varied from 45 to 65 years, where the mean age was 57.11 years old. The age of females ranged from the minimum of 40 to the maximum of 63 years old. The mean age for female respondents was 59.45 years old.

3.2 *Cronbach Reliability Coefficients*

Stress Indicator Questionnaire (SIQ), which consists of physical, sleep, behavioural, emotional, and personal habits, was used in this study. Cronbach alpha was found ranging from 0.86 to 0.88 (Table 2).

Table 3 Stress mean of obese and nonobese

Stress indicator	Mean	
	Obese	Nonobese
Physical	20.7343	12.3172
Sleep	19.2987	17.7121
Behavioural	21.5171	16.3917
Emotional	23.3312	14.0010
Personal	15.4591	11.3261

Table 4 Level of stress

Subject	Mean	*t*-Value	*P*-Value
Obese	20.0681	13.3120**	0.000
Nonobese	14.3496		

** $p < .01$

3.3 Mean of Stress Indicator Between Obese and Nonobese

Stress indicator items were evaluated, emotional symptoms have the highest mean ($\bar{x} = 23.3312$) on obese, while sleep symptoms ($\bar{x} = 17.7121$).

3.4 Level of Stress Between Obese and Nonobese

Table 3 shows the *t* test scores for the level of stress between obese and nonobese, $t(121) = 13.3120$, $p < .01$ (Table 4).

4 Discussion

Comparison of mean between obese and nonobese showed that all the symptoms were higher in obese compared to nonobese. This showed that the obese population has a tendency of high-level stress in all the stress symptoms. The physical symptoms associated with stress include the experience of nervousness, sweaty palms, chronic headaches, muscle tension, and feeling of butterflies in the stomach. The sleep symptoms of stress include take pills to get to sleep, experience nightmares, and tiredness. The behavioural symptoms of stress include drinking alcohol, use drugs, and eat while doing activities. The emotional symptoms of stress include lack of relax, worry, and emotion not stable. The personal habit includes religious belief, lack of time for entertainment and could not release stress.

Food and drink are necessary to sustain life. However, for the vast majority of people, interaction with food is pleasurable and a source of relief from daily stress [3].

The level of physical, behavioural, and emotional symptoms of stress is perceived to lower by eating large amount of food. Research on eating behaviour showed that people tend to be calm and less emotional after overeating. People are typically alert and irritable when hungry and calm and sleepy when full [5]. According to Steiner [16], pleasure gain from eating food positively affect emotional. This study gains support from the research done by Pecoraro et al. [15] on rats showed a significant decreased in the level of stress hormone immediately after consuming food. Therefore, most probably the present research indicated that obese people use eating as a coping strategy to reduce their stress level. In other words, whenever obsess got into stress, they eat frequently or in large amount to reduce the level of stress.

The majority of early work on stress–eating relationships investigates the General Effect Model [3]. This theory assumed that stress produces a general response to eating. In other words, stress produces physiological changes in the organism and these change the eating behaviour. Research in this paradigm has particularly focused on animal research. In an experimental study, rats have been pinching their tails and observed significant increases in licking and eating food [2]. Tail pinching in this experiment was considered as represent stress-induced eating. In other studies, experiments were also done using electric shock and cold water as stressors. These studies also support the General Effect Model [3]. The General Effect Model principally assumes that stress produces a physiological or biological change that in turn causes a change in eating.

Moreover, a few studies indicated that overweight individuals were more likely to respond stressful or emotional stimuli by eating. Out of 11 studies reviewed by Greeno and Wing [7], three demonstrated an increase in eating for obese individuals when stress compared to nonobese. However, five further studies failed to find a relationship between stress and eating in obsess individuals. Stress is assumed to lead to increased eating in emotional eaters because they fail to distinguish between hunger and anxiety [3].

5 Conclusion

The present study showed that the symptoms of stress include physical, sleep, behavioural, emotional, and personal habit, were higher on obese compared to nonobese respondents. It is recommended for obese Malaysian to take balanced diet, as shown in this diagram. For every 3500 kcal of excess energy accumulated, 1 lb (0.45 kg) of fat is stored in the body. Therefore, consuming less fat meal or healthy daily diet is very important for obese (Fig. 3).

Most probably obese population should use other strategies to reduce their stress level than food. If obese could not avoid eating large amount of food to reduce their stress, they should replace it with low-calorie or vegetarian food. In fact, diets rich vegetables and fruits lower blood pressure and decrease the risk of heart attack, stroke and certain types of cancers. Besides that, obese individuals should take part

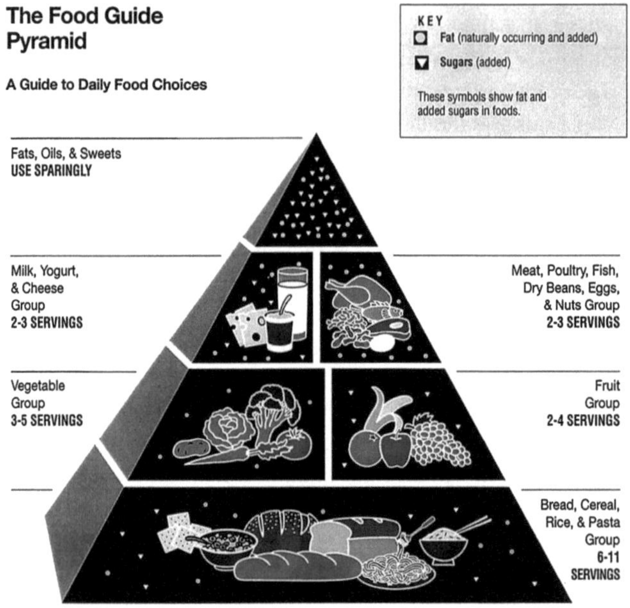

Fig. 3 The recommended healthy daily diet

in physical activity, not only to reduce their level of stress but also to prevent disease related to obese. Physical activities tend to increase metabolic rate, which is very important to cut down the accumulated excessive calories in the body.

References

1. Ampofo Boateng, K. (2009). *UnderstandingSport Psychology*. Selangor, Malaysia: UiTM Press.
2. Antelman, S. M., Szechtman, H., Chin, P., & Fisher, A. E. (1975). Tail Pinch-Induced eating, gnawing and licking behaviour in rats: Dependence on the Nigrostriatal Dopamine System. *Brain Research, 99*, 317–319.
3. Corner, M., & Armitage, C. J. (2002). *The Social Psychology of Food*. Buckingham: Open University Press.
4. Epel, E., Jimenez, S., Brownell, K., Stroud, L., Stoney, C., & Niaura, R. A. Y. (2004). Are stress eaters at risk for the metabolic syndrome? *Annals of the New York Academy of Sciences, 1032*(1), 208–210.
5. Gibson, E. L. (2006). Emotional influences on food choice: Sensory, physiological and psychological pathways. *Physiology & Behavior, 89*(1), 53–61.
6. Gilbert, S. (1989). Psychological Aspect of Obesity and its Treatment. In R. Stepherd (Ed.), *Hand of the Physiology of Human Eating*. Chichester: Wiley.
7. Greeno, C. G., & Wing, R. R. (1994). Stress-induces eating. *Psychological Bulletin, 115*, 444–464.
8. Han, K., Trinkoff, A. M., Storr, C. L., & Geiger-Brown, J. (2011). Job stress and work schedules in relation to nurse obesity. *Journal of Nursing Administration, 41*(11), 488–495.

9. Hussien, T., & Hussien, S. (2006). *Strategies for coping Educational and Psychological stress*. Daf Alfiker: Amiman.
10. Jeffery, R. W., Drewnowski, A., & Epstein, L. H. (2000). Long-term maintenance of weight loss: Current status. *Health Psychology, 19*, 5–16.
11. Kolata, G. (1985). Obesity declared a disease. *Science, 227*, 1019–1020.
12. Leon, G. R., Fulkerson, J. A., Perry, C. L., & Early-Zald, M. B. (1995). Prospective analysis of personality and behavioral vulnerabilities and gender influences in the later development of disordered eating. *Journal of Abnormal Psychology, 104*(1), 140–149.
13. Medical-Dictionary, 2013. https://medicaldictionary.thefreedictionary.com/obesity
14. National Heart, Lung and Blood Institute. (1998). U.S Department of Health and Human Service. (http://www.nhlbi.nih.gov/news/press-releases/1998/first-federal-obesity-clinical-guidelines-released).
15. Pecoraro, N., Reyes, F., Gomez, F., Bhargava, A., & Dallman, M. F. (2004). Chronic stress promotes palatable feeding, which reduces signs of stress: Feed forward and feedback effects of chronic stress. *Endocrinology, 145*(8), 3754–3762.
16. Steiner, J. E. (1974). Discussion paper: Innate, Discriminative Human facial expressions to taste and smell stimulation. *Annals of the New York Academy of Sciences, 237*, 229–233.
17. Thompson, D. (2009). *The Link Between Stress and Obesity*. Everyday Health. http://www.everydayhealth.com/diet-nutrition/food-and-mood/stress-and-dieting/stress-and-other-causes-of-obesity.aspx.
18. Van-Strien, T., & Ouwens, M. A. (2003). Counterregulation in female obese emotional eaters: Schachter, Goldman, and Gordon's (1968) Test of Psychosomatic Theory Revisited. *Eating Behaviors, 3*(4), 329–340.
19. WHO. (2003). Global Database on Body Mass Index. http://apps.who.int/bmi/index.jsp?introPage=intro_3.html.

Relationship Between Launch Angle and Ball Distribution of Kuda Kick in Sepak Takraw Service

Shariman Ismadi Ismail, Nabilah Husna Mohd Zani and Norasrudin Sulaiman

Abstract Service is an important move in the sepak takraw sport. A study to observe the relationship between the service ball's launch angle and ball distribution in the opponent's area in real sepak takraw match situations was carried out. Video analyses were performed on 10 experienced national university level servers (age = 21.61 ± 2.0 years; height = 1.71 ± 0.04 m; mass = 73.70 ± 10.17 kg). All videos were recorded during an inter-university sports tournament, starting from the group match, until the final match of the competition. The analyses focused only on the 'Kuda Kick' service technique. In order to analyze ball distribution during serve, the serve's receiving side of the court was categorized into nine zones (zone A–I). A high definition video camera was utilized to record the entire two-sided court area from a sagittal plane view. The serve launch angles were then obtained from video analysis. It was identified that there were 7 ranges of launch angles during service. The results indicated that the highest ball distribution is in zone F (26.6%), followed by zone B (24.8%) and zone E (17.8%); whereas the lowest ball distribution is in zone G (1.0%). In terms of the serve launch angle, the highest distribution is at a range of 1–10°(42.2%), and 11–20°(36.0%). Based on these 2 launch angle ranges, range 1–10° recorded 31.27% ball distribution in zone B, and 24.17% in zone F. The launch angle of 11–20° was mostly distributed in zone F (32.22%), followed by zone B 20.56%. The results indicated that high ball distribution area created a 'T-Zone' in the court area. This study also shows that there was a moderate correlation between serve launch angle and ball distribution area ($r = 0.374$; $p < 0.05$).

Keywords Video analysis · Sepak takraw · Launch angle · Service

---

S. I. Ismail (✉) · N. H. Mohd Zani · N. Sulaiman
Faculty of Sports Science and Recreation, Universiti Teknologi MARA, 40450 Shah Alam, Selangor, Malaysia
e-mail: shariman_ismadi@salam.uitm.edu.my

N. Sulaiman et al. (eds.), *Proceedings of the 3rd International Colloquium on Sports Science, Exercise, Engineering and Technology*,
https://doi.org/10.1007/978-981-10-6772-3_9

1 Introduction

Sepak takraw is a kicking sport that is categorized as a difficult net barrier attributable to its essential acrobatic twists, quick reflexes, turns or flip, and agile human body ability. Sepak takraw is a fast-moving sport that requires skills in running, spiking, kicking, and passing. Serving movement in sepak takraw is considered the most important move as a point-gainer [1]. The sepak takraw player must jump parallel or higher than a net that is 5 feet in height so that they can make a sepak sila service, or even block the ball.

The typical serves that are performed in the takraw game are Kuda Kick service and Sila Kick service. The Kuda Kick service is performed by using the instep of the foot, and Sila Kick service is performed by using the inside of foot [1]. Sila Kick service is a traditional serve that is used before Kuda Kick serve was introduced during the 18th SEA Games in Chiang Mai in 1995. They are nowadays applied by most elite sepak takraw athletes.

Trajectory can be described as the path that a moving object follows [2]. The objects are thrown horizontally. Every server has a different trajectory of the ball during serve the ball. There are a few factors that influence the outcome of the ball, namely, lower body power and strength, body fitness, current and past injury, level of concentration, level of serve a skill, angle of hip flexion during performed serve, and foot-to-foot distance. The trajectories of ball and players express tactic information; the trajectory analyses are essential for the sepak takraw tactical analysis [3]. This study focuses on one of the sepak takraw ball trajectories component during Kuda Kick serve: the launch angle of the ball.

2 Method

2.1 Participants

The movement of 10 experienced national university level servers was recorded (age $= 21.61 \pm 2.0$ years; height $= 1.71 \pm 0.04$ m; mass $= 73.70 \pm 10.17$ kg). All videos were recorded during an inter-university sports tournament, starting from the group match, until the final match of the competition

2.2 Instrumentation

The analyses focused only on the Kuda Kick service technique. A high-definition video camera at 50 fps was located 7.5 m from the center of the court, to record the entire two-sided court area from a sagittal plane view. The camera height was set at

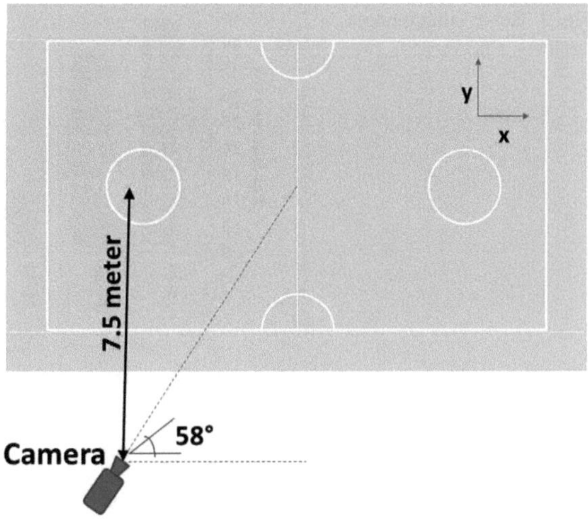

Fig. 1 Camera setting

3 m [4]. The camera was positioned at a 58° slant to obtain an optical intersection at the center of the court (Fig. 1). The camera setting parameters were set to obtain a calibrated region of the court size [4].

2.3 *Procedures and Analysis*

In order to analyze ball distribution during a serve, the serve's receiving side of the court was categorized into nine zones (zone A–I), as shown in Fig. 2.

The serve launch angles were then obtained from video analysis using Kinovea Motion Analysis Software (v0.8). The launch angles were then categorized into nine ranges of angles: 1 (0° to 10°), 2 (11° to 20°), 3 (21° to 30°), 4 (31° to 40°), 5 (41° to 50°), 6 (51° to 60°), 7 (−1° to −10°), 8 (−11° to −20°) and 9 (−21° to −10°).

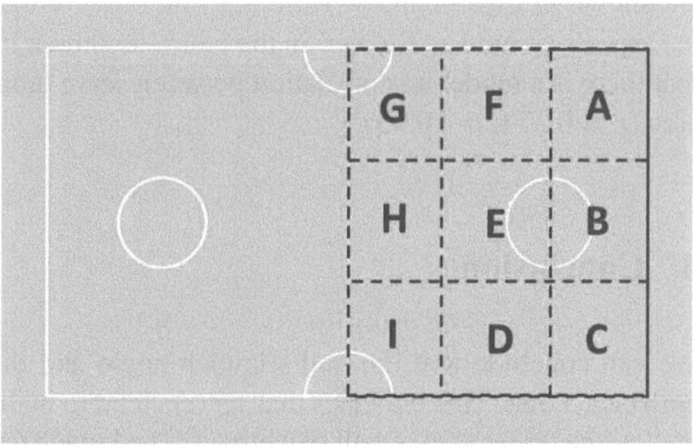

Fig. 2 Receiving side court zone

Fig. 3 Serve launch angle ranges

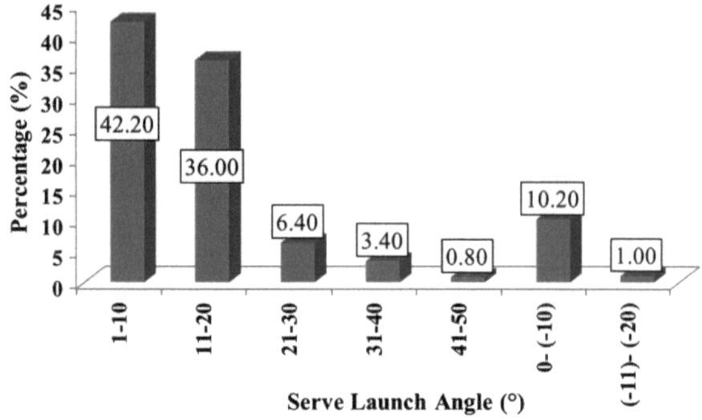

Fig. 4 Ball distribution zone

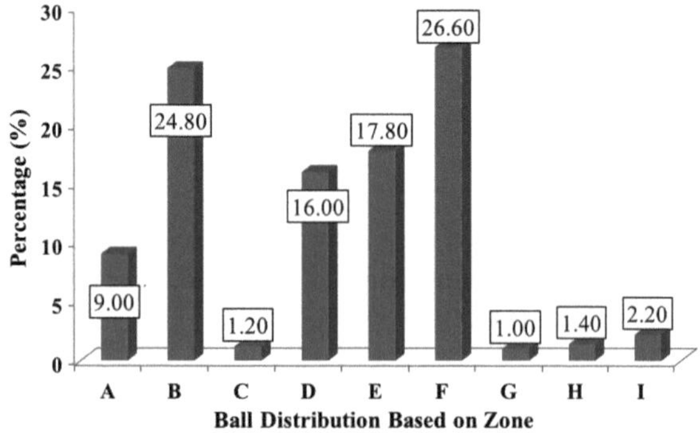

3 Results and Discussion

The results indicated that the highest ball distribution is in zone F (26.6%), followed by zone B (24.8%) and zone E (17.8%); whereas the lowest ball distribution is in zone G (1.0%). In terms of the serve launch angle, the highest distribution is at range 1 (42.2%) and range 2 (36.0%). Based on these 2 launch angle ranges, range 1 recorded 31.27% ball distribution in zone B, and 24.17% in zone F. Launch range 2 is mostly distributed in zone F (32.22%), followed by zone B (20.56%). The overall results are shown in Figs. 3 and 4. Here, we can observe that the highly targeted zone by the servers created a 'T-Zone' in the court, as shown in Fig. 5. The study also shows that there is a moderate correlation between serve launch angle and ball distribution area ($r = 0.374$; $p < 0.05$).

4 Conclusion

We can conclude that the ball's launch angle and distribution area are correlated with each other. This indicates that servers tend to maintain their serve parameters as well as the target area of ball placement. The launch angle of the ball gives a specific

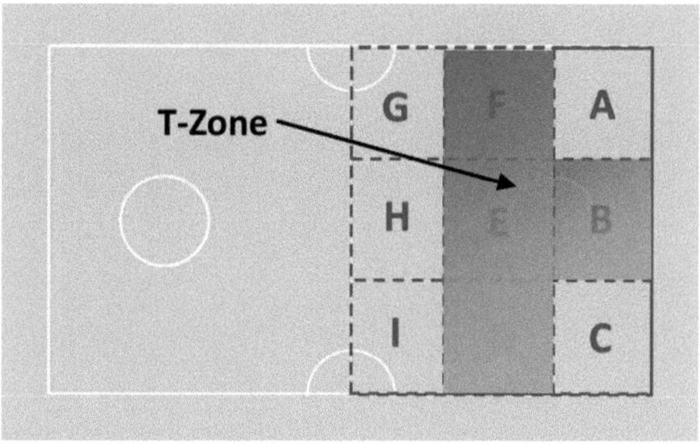

Fig. 5 T-Zone: high ball distribution area

court zone for the ball to be distributed, as shown based on the T-Zone. However, this study did not consider other ball trajectories components such as ball velocity and the foot-to-ball contact's relative height. Therefore, further analyses that integrate with other trajectories components must be performed in order to better understand this study. The combined information of the opponent ball's launch angle and velocities behavior could help a sepak takraw team to make improvement in their strategies and game plans.

Acknowledgements This work was sponsored by Universiti Teknologi MARA [600-RMI/DANA 5/3/ARAS (54/2015)]. We also would like to thank the Faculty of Sports Science and Recreation, Universiti Teknologi MARA, Malaysia for supporting this study.

References

1. Usman, J., & Abas, W. A. B. W. (2004). A comparative analysis on selected kinematics parameters between the "sepak kuda" serve and the "sepak sila" serve in sepak takraw. In *Proceedings of the XXIInd International Symposium on Biomechanics in Sports* (pp. 322–325). Ottawa.
2. Ranaware, D., & Nannikar, A. (2013). Mechanism of projectile motion. In *Proceedings of International Conference on Technology and Business Management* (pp. 1243–1249).
3. Niu, Z., Gao, X., & Tian, Q. (2012). Tactic analysis based on real-world ball trajectory in soccer video. *Pattern Recognition, 45,* 1937–1947.
4. Vander Linden, D. W., Carlson, S. J., & Hubbard, R. L. (1992). Reproducibility and accuracy of angle measurements obtained under static conditions with the motion analysis™ video system. *Physical Therapy, 72,* 300–305.

A Study of New Regime Interval Training Exercise on Obesity Management Among Sedentary Overweight Working Women

Mastura Johar, Rogemah Ramli, Rozita Abd Latif and Ahmad Termizi

Abstract The purpose of this investigation is to evaluate the effect of 12 weeks of a new regime interval training exercise, an intervention in the treatment of obesity diagnosis among overweight sedentary working women in Universiti Tenaga Nasional. The new regime interval training exercise program is a combination of aerobic dance activity, circuit training, and interval training which improvise own body weight. It combines physical training that includes alternating of low and medium intensity of exercise workout which involves a physical conditioning training with medium volume and low resistant of training with a short rest time suitable for sedentary and overweight individuals. Forty subjects between the ages of 25–55 years old have had a diagnosis as sedentary, overweighted, and had a fitness score below the mean value. Subjects were randomly assigned to 12 weeks of either new regime interval training exercise program as a treatment group or an aerobics dance exercise as the control group. A pretest–posttest control research design was utilized. Both groups met for 50 min, three times per week for a total of 36 sessions. Statistical analysis includes (group x time) paired t-test and independent sample t-test was used to determine between and within group mean differences. Results suggested that subjects in the treatment of new regime interval training experienced positive improvement in obesity diagnosis after 12 weeks of treatment within the group effect $p < .05$. Results for obesity diagnosis were, BMI (mean pre: 30.57, mean post: 29.24), PBF from (mean pre: 43.52, mean post: 43.00), WHR (mean pre: .928, mean post: .924), and SMM (mean pre: 23.23, mean post: 23.32). Therefore, it was concluded that the intervention of a new regime interval training exercise exercise program had improved obesity diagnosis, after 12 weeks of treatment and contributed to positive findings among overweight sedentary women in higher education organization in Malaysia.

M. Johar (✉) · R. Ramli
College of Engineering, Universiti Tenaga Nasional, Kajang, Malaysia
e-mail: mastura@uniten.edu.my

R. A. Latif
Sport Science and Recreation Department, Universiti Teknologi Mara, Shah Alam, Malaysia

A. Termizi
College of Graduate Studies, Universiti Tenaga Nasional, Kajang, Malaysia

N. Sulaiman et al. (eds.), *Proceedings of the 3rd International Colloquium on Sports Science, Exercise, Engineering and Technology*,
https://doi.org/10.1007/978-981-10-6772-3_10

Keywords New regime interval training · Obesity · Circuit training · Aerobics dance

1 Introduction

Nowadays, obesity issues are faced by adults associated with physical inactivity and unhealthy diet, resulting in high level of stress and poor physical health [1]. Obesity is considered as worldwide pandemic. It has become a major health problem in some developed countries such as the United State of America and Great Britain, [2] also poor country such as South Africa [3]. Further, developing country such as Malaysia is facing the obesity problem too [4]. According to the Ministry of Health in 2006, about 70% of Malaysian adults are not physically active and practice sedentary lifestyle [5]. Obesity contributed to a negative effect on individual appearance and attractiveness to both man and women [6]. Malaysia Ministry of Health [7] has been seriously considered specifically on non-communication disease (NCD) such as cardiac disease, cancer and diabetes are occur due to the unhealthy lifestyle leading to obesity and overweight problems. Obesity is when an individual weight is 20% higher than their recommended weight. In 2006, about 43% of Malaysian adults were categorized as overweight and obese which indicates that Malaysians include passive and non-active citizens [8]. The modern lifestyle with modern technology facilities had turned many individuals become inactive thus; it leads to the overweight and obesity. Obesity problem occurs due to an imbalance in the energy intake and expenditure by the individuals. It is proven that excessive energy intake will lead to the formation of fatty tissues that cause obesity. Obesity can be classified into several parts according to body mass index (BMI); BMI of 25.0–29.9 is categorized as overweight, BMI of 30–34.9 is categorized as Obese Class 1, BMI of 35–39.9 is categorized as Obese Class 2 and BMI of 40 and above is categorized as Obese Class 3 [9]. Current obesity issue in Malaysia is critical as many Malaysian individuals who are focusing on job and trying to comply on archiving their employer and company target. They have less time to rest, sleep, and practice balance diet in their daily life, while in free time they spend their time by watching television, using smart phone application and playing computer game [10], instead of involving in physical activity

A large amount of money have been spent on health care in the regards to this obesity problem. There are many types and designs of training programs which have been created by physical trainers and recommended to the obese population in order to reduce the rate of obesity in Malaysia. Disappointingly, these kinds of protocols have led to negligible weight loss [11, 12]. Thus, exercise protocols that can be carried out by overweight, inactive individuals that more effectively reduce body fat are required. Accumulating evidence suggests that a combination of aerobic dance activity, circuit, and interval training that improvise own body weight during the exercise program has the potential to be an economical and effective exercise protocol for reducing fat of overweight individuals. Therefore, the purpose of this investigation is to evaluate the effects of a 12 weeks of a new regime interval training exercise intervention in the treatment of obesity diagnosis among sedentary overweight women. Most exercise protocols designed to induce fat loss have focused on regular steady-state exercise such as walking and jogging at a moderate intensity. For optimum health and fitness target in exercise program, it should involve both aerobic and resistance training exercises [13]. A research by [14] showed that by doing aerobic exercise regularly, there will be a positive influence on mental health and reduces the affecting states. A study by [15] showed that aerobic exercise will increase satiety compared to resistance exercises. Other than that, aerobic dance also had been proved to improve neuro-cognitive functions for the elderly and exercise with coordination can also improve attention ability [16].

2 Method

The study is to observe any beneficial effects of the new regime interval training exercise on 20 subjects experienced sedentary lifestyle and overweight. The research design used in this study is a quasi-experimental design, pretest and posttest control group designs. Quasi-experiment design provides as much control as possible [17, 18]. One of the strongest and most widely used quasi-experimental design, which will differ from other experimental design because treatment and control groups are not equivalent. Comparing pretest results will indicate degree of equivalency between treatment and control group [19]. This experimental research attempts to manipulate the independent variable and study the effect of this manipulation or treatment on the dependent variable [20]. Obesity was diagnose measuring four parameters namely: body mass index (BMI), waist-hip ratio (WHR), percent body fat (PBF), and skeletal muscle mass (SMM). A pretest was administered to all subjects of this study to measure and diagnose obesity parameters before treatment condition was introduced. However, a takeoff pretest was administered to gauge subjects with having sedentary lifestyle and overweight. Subjects were randomly assigned to a 12-week Baton Dance exercise intervention or aerobics dance exercise as the control group. A pretest–posttest control research design was utilized. 20 subjects were measured again at week 12 after the treatment is given [21]. INBODY machine measure obesity diagnosis. Subjects were required to participate in 36 sessions, for 60 min each session, 3 times per week for 12 weeks period working at 60% to 70% of maximum training heart rate. It is gauged by comparing the differences between pretest and posttest scores of the intervention group with that of the control group. Their level of both measurements (obesity) was measure at posttest after the 36 session at week 12 to determine the effect of the intervention program on the dependent variables. Figure 1 shows the treatment assignment and the pretest–posttest design used in this research study.

QUASI-EXPERIMENTAL DESIGN, PRE-TEST AND POST-TEST CONTROL GROUP DESIGNS			
GROUP 1 (TREATMENT) (n=20)	**GROUP 2 (CONTROL) (n=20)**	**PRE-TEST** Week 1 Session 1	**POST-TEST** Week 12 Sessions 36
Type: New regime interval training exercise	**Type:** Aerobics Dance Exercise	**Measure 1**	**Measure 2**
Training Heart Rate: 60-70% THR	**Training Heart Rate:** 60-70% THR	**1. Obesity Diagnosis** *(BMI,SMM,WHR & PBF)* **2. Demographic Information** *(Age, Occupation, Marital Status, Academic, Qualification)* **3. Par-Q Test – Health Stage** **2. Attendance** **4. Consent** **5. Attendance**	
Music Tempo Range: 120-130 BPM	**Music Tempo Range:** 135 -145 BPM		

Fig. 1 Research design

3 Results

3.1 Descriptive Data Analysis

The results obtained from the preliminary analysis of the frequency distribution are shown in Table 1. Twenty subjects were assigned in the treatment group and 20 were in the control group. Before the statistical analysis was done, the respondent profile data was examined. As depicted in Table 1. 20% subjects were married, 12% were not married 8% were divorced. Age group was categorized according to three ranges where results indicate that age range 25–35, 30%, age 36–45 were the majority subjects, 45% and age range of 46–55, 25%. From the total number of subjects involved in this research, 30% were Academician/Managerial, 35% were Administrator, and 35% were Support Staff. Subjects with Ph.D./Masters were 27.5%, Degree was 20%, Diploma was 25.0%, and SPM/MCE was 27.5%. Majority of the subjects were Malay 77.5%, while only 15% were Chinese and 7.5% were Indians. In addition, since the analysis of the comparison involved two independent samples, the normality of the data distribution and equality of group variance were assessed before the parametric statistics were used in the analysis. Whether or not the normality of a distribution is rejected will depend on the ratio of skewness and kurtosis to its statistic standard error. Results show that all the dependent variables that were examined and the ratio of skewness and kurtosis on their respective statistic was within the range of −2.00 to + 2.00 (refer to Table 1). Therefore, parametric statistical analyses were used in this study.

Table 1 Frequency distribution of subject is age category, marital status, job category and academic level, and races

Variables	Frequency	%	Skewness	Kutosis
Age (years)				
25–35	17	29.3	.550	1.06
36–45	15	25.9	.580	1.12
46–55	8	13.8	.752	1.48
Marital status				
Single	10	17.2	.687	1.33
Married	25	43.1	.464	.902
Divorce	5	8.6	.913	2.00
Job category				
Academician	14	24.1	.501	.972
Administrator	3	5.2	–	–
Support staff	21	36.2	.597	1.15
Others	2	3.4	–	–
Academic level				
Ph.D. Masters	12	20.7	.913	2.00
Degree	11	19	.637	1.23
Diploma	12	20.7	.661	1.27
SPM/MCE	5	8.6	.637	1.23
Race				
Malay	31	77.5		
Chinese	6	15		
Indian	3	7.5		

3.2 *Statistic Data Analysis Obesity Diagnosis*

In response to the research question:

To investigate the difference between the experiment (New regime interval training Exercise) and control group (aerobics dance) in terms of the difference between pretest and posttest scores on obesity diagnosis namely BMI, PBF, WHR, and SMM.

A paired t-test was conducted to evaluate the impact of the intervention on participant obesity diagnosis. There was a statistical significant improvement in all obesity diagnosis except for SMM. There was a statistical significant decrease in BMI during pretest ($M = 30.57$, SD = 5.04) to posttest ($M = 29.82$, SD = 5.03) t (39) = 1.17, $p < .05$. There was a statistical significant decrease in PBF during pretest ($M = 4.352$, SD = 5.17) to posttest ($M = 43.00$, SD = 5.55) t (39) = .984, $p < .05$. There was a statistical significant improvement in WHR during pretest ($M = .928$, SD = .051) to posttest ($M = .924$, SD = .052) t (39) = .0073, $p < .05$. However, there was no

Table 2 Results of paired t-test during pretest and posttest on obesity diagnosis

Obesity diagnosis	N	Mean		SD	t	df
Body Mass Index (BMI)	40	Pre	30.572	5.0426	4.402	3
		Post	29.243	5.0648		
Percent Body Fat (PBF)	10	Pre	43.525	5.1732	.984	39
		Post	43.000	5.5524		
Waist-Hip Ratio (WHR)	40	Pre	.9287	.05195	.07	39
		Post	.9245	.05238		
Skeletal Muscle Mass (SMM)	40	Pre	23.235	3.045	−.48	39
		Post	23.320	2.730		

statistical significant improvement in SMM during pretest ($M = 23.23$, SD = 3.045) to posttest ($M = 23.32$, SD = 2.73) t (39) = .486, $p > .05$. Table 2 described the results clearly.

Further to that, comparisons between the two groups were made using an independent- sample t-test to compare the obesity diagnosis namely BMI, PBF, WHR, and SMM scores during week 1 and week 12 between treatment and control groups. Results indicate during week 1, there was no significant difference in all the obesity diagnosis variables for treatment and control groups. BMI ($M = 30.47$, SD = 4.568), and control group ($M = 30.68$, SD = 5.59); t (38) = −.127, $p > .05$, PBF ($M = 43.40$, SD = 5.37), and control group ($M = 43.65$, SD = 5.098); t (38) = −.151, $p > .05$. WHR ($M = .928$, SD = .0492), and control group ($M = .929$, SD = .0558); t (38) = −.0306, $p > .05$ and SMM ($M = 23.30$, SD = 3.031), and control group ($M = 23.17$, SD = 3.136); t (38) = .133, $p > .05$. Further to that, at week 12 during posttest, it was reported that there were no significant differences in all scores for both the treatment group and control group. BMI ($M = 29.12$, SD = 4.21), and control group ($M = 30.52$, SD = 5.76); t (38) = −.873, $p > .05$, PBF ($M = 42.80$, SD = 5.37), and control group ($M = 43.20$, SD = 5.301); t (38) = −.225, $p > .05$. WHR ($M = .923$, SD = .0489), and control group ($M = .926$, SD = .0568); t (38) = −.119, $p > .05$ and SMM ($M = 23.64$, SD = 2.827), and control group ($M = 23.01$, SD = 2.664); t (38) = .725, $p > .05$. Table 3 described the results clearly.

3.3 *Discussion on Obesity Diagnosis Effect*

This study produced results which corroborated the findings of a great deal of the previous work in this field [22, 23]. A study by [19] examined the effects of aerobics exercise and interval training on body mass index on two different 15-week treatment on various fitness parameters in females. However, results of the present study did not show a significant difference between the treatment and control groups. Somehow both groups improved in all the obesity diagnosis namely: BMI, WHR, and

Table 3 Independent—sample t-test results on obesity diagnosis scores between treatment and control groups

Var		N	M	SD	t	df	p	Mean differ
Pretest								
BMI	Treat	20	30.4	4.568	−.12	38	.130	−.205
	Cont	20	30.6	5.594				
PBF	Treat	20	43.4	5.376	−.15	38	.971	−.250
	Cont	20	43.6	5.098				
WHR	Treat	20	.928	.0492.05	.03	38	.464	−.0005
	Cont	20	.929	58				
SMM	Treat	20	23.3	3.031	.13	38	.820	−1.395
	Cont	20	23.1	3.136				
Posttest								
BMI	Treat	20	29.1	4.21	.87	38	.074	−1.395
	Cont	20	30.5	5.76				
PBF	Treat	20	42.8	5.923	−.22	38	725	−.400
	Cont	20	43.2	5.301				
WHR	Treat	20	.923	.0489	−.11	38	.342	−.002
	Cont	20	.925	.0568				
SMM	Treat	20	23.6	2.827	.72	38	.992	.1300
	Cont	20	23.0	2.664				

PBF except for SMM after the treatment. It seems possible that these results are due to both the intervention group and control group having experienced aerobics dance program but with a different approach, intensity and impact. These findings further support the idea of the thesis by [24] Dance for Health, an intervention program designed to provide an enjoyable exercise program for African American and Hispanic adolescents. It had a significant effect on improving aerobics capacity, helping students maintain or decrease weight, and improving attitudes towards physical activity and physical fitness. In the first year of the program (1990–91), approximately 110 boys and girls aged 10–13 years participated in an aerobics dance pilot program three times per week for 12 weeks. Dance for Health was revised and continued in the 1992–93 school years with seventh grade students and an added culturally sensitive health curriculum. Forty-three students were randomly delayed to Dance for Health and 38 to usual physical activity. Those in the intervention class received a health education curriculum twice a week and a dance-oriented physical education class three times a week. The usual physical activity consisted mostly of playground activities. Students in the intervention had a significantly greater lowering in body mass index and resting heart rate than students in regular physical activity and the results supported the present study.

The findings of the present study are also important and supported [22] examined on the short and long-term changes in BMI and body composition, produced by diet combined with structured aerobics exercise. Sixteen-week randomized controlled trial with a 1-year follow-up, forty obese women (mean body mass index 32.9 kg/m^2; mean weight, 89.2 kg) with a mean age of 42.9 years (range, 21–60 years) improved their BMI with a structured aerobics exercise or with the combination of low-fat diet of about 1200 kcal.

4 Conclusion

The findings of the study revealed that subjects of sedentary and overweight women in Universiti Tenaga Nasional showed statistically significant main effects on the dependent variables on obesity diagnosis $p < .05$, after undergoing a treatment of baton dance (treatment group) and aerobics dance (control group) for 12 weeks. In addition, the obesity diagnosis on all variables experience significant difference except for SMM variable, examination suggests that there was not statistically significant main effect on dependent variable of SMM variables during pretest to posttest between treatment and control group. Results of independent t-test suggest that there was no significant difference in both treatment and control groups on the obesity diagnosis, this interaction effect indicates that the difference between the experiment and control group on the linear combination of all dependent variables were different at pretest (week 1) than it is at posttest (week 12). Examination of the means suggests that this is because groups do not differ on either dependent variable at the time of the pretest, but they do differ, particularly on all the dependent variables at the time of the posttest (week 12) significantly. In addition, examination of the means suggests that there was a change in the posttest (week 12) outcomes held for both the treatment and control groups. Therefore, the new regime interval training exercise seems to be an efficient means of exercise to help decrease percent body fat (PBF), waist-hip ratio (WHR), and body mass index (BMI). Traditionally, resistance training often is performed separately from aerobic training typically on two or three nonconsecutive days each week. The American College of Sports Medicine (ACSM) recommends 8–12 repetitions of a resistance training exercise for each major muscle group at an intensity of 40–80% of a one-repetition max (RM) depending on the training level of the participant. Two to three minutes of rest is recommended between exercise sets to allow for proper recovery. Two to four sets are recommended for each muscle group. While standard guidelines for aerobic training recommend 150 min per week of moderate intensity exercise for 30–60 min per session and/or 75 min per week of vigorous intensity exercise (64–90% $\dot{V}$O2max) for 20–60 min per session [25].

In conclusion, the new regime interval training exercise produces significant increases in aerobic and anaerobic fitness and brings about significant skeletal muscle adaptations that are oxidative and glycolytic in nature. The effects of the new regime interval training exercise on subcutaneous and fat loss are promising but more studies using overweight individuals need to be carried out. Given that the major reason given for not exercising is time [26], it is likely that the brevity of the new regime interval training exercise protocols should be appealing to most individuals interested in fat reduction. The optimal intensity and length of the circuit and rest periods together with examination of the benefits of other the new regime interval training exercise modalities need to be established. The new regime interval training exercise has provided an effective and efficient program for sedentary and obese individual, and it uses a program with a combination of aerobic dance activity, circuit, and interval training that improvise own body weight during the exercise program. Individuals who were previously sedentary and overweight believed that they did not have the time for exercise can now trade total exercise time for total exercise effort and get similar or better health and fitness benefits. Recovery is the best indication of fitness, and the new regime interval training exercise which is the combination of aerobics, circuit and interval dramatically improve recovery and the new regime interval training exercise exercise has improved the sedentary and overweight individuals physiologically and psychologically well-being and lifestyle.

References

1. American College of Sports Medicine. (2001). *ACSM'S resource manual for guidelines for exercise testing and prescription* (6th ed.). Philadelphia: Lippincott Williams and Wilkins.
2. Kinge, J. M., & Morris, S. (2013). Variation in the relationship between BMI and survival by socioeconomic status in Great Britain. *Economics and Human Biology, 12,* 67.
3. Case, A., & Menendez, A. (2009). Sex differences in obesity rates in poor countries: Evidence from South Africa. *Economics and Human Biology, 7,* 271.
4. Wang, S. L. A. N. S. Y. (2014). Decomposing race and gender differences in underweight and obesity in South Africa. *Economics and Human Biology, 15,* 23.
5. Ministry of Health. (2003). *Seminar on findings of The Malaysian Adult Nutrition Survey (MANS) 2003.* Putrajaya: Family Health Development Division, Ministry of Health Malaysia.
6. Fletcher, J. M. (2014). The interplay between gender, race and weight status: Self perceptions and social consequences. *Economics and Human Biology, 14,* 79.
7. MOH. (2011). *Annual report ministry of health 2011* (pp. 1–351).
8. Mansor, M., & Harun, N. Z. (2014). Health issues and awareness, and the significant of green space for health promotion in Malaysia. *Procedia-Social and Behavioral Sciences, 153,* 209.
9. https://medlineplus.gov/ency/article/007196.htm.
10. Saelens, B. E., Sailis, J. F., Nader, P. R., Broyles, S. L., Berry, C. C., & Taras, H. L. (2002). Home environmental influences on children's television watching from early to middle childhood. *Journal of Developmental and Behavioral Pediatrics, 23,* 127–132.
11. Shaw, K., Gennet, H., O'Rourke, P., Del Mar, C. 2006. *Exercise for overweight or obesity.* Wiley, The Cochrane Collaboration.
12. Wu, T., Gao, X., Chen, M., & Van, Dam R. M. (2009). Long-term effectiveness of diet-plus-exercise interventions vs diet-only interventions for weight loss: A meta-analysis: Obesity management. *Obesity Reviews, 10*(3), 313–323.
13. *Human Kinetics by making exercise your medicine*. IL, U.S.A.: Human Kinetics.
14. Guelfi, K. J., Donges, C. E., & Duffield, R. (2012). Beneficial effects of 12 weeks of aerobic compared with resistance exercise training perceived appetite in previously sedentary overweight and obese men. *Metabolism, Clinical and Experimental, 62,* 235–243.
15. Petruzzello, S. J. (1995). Anxiety reduction following exercise: Methodological artifact or "real" phenomenon? *Journal of Sport and Exercise Psychology, 17,* 105–111.
16. Kimura, K., & Hozumi, N. (2012). Investigating the acute effect of an aerobic dance exercise program on neuro-cognitive function in the ederly. *Psychology of Sport and Exercise, 13,* 623–629.

17. Creswell, John W. (2002). *Research design: Qualitative, quantitative, and mixed methods approaches*. Thousand Oaks, CA: Sage Publications.
18. Shadish, W. R., Cook, Thomas D., & Campbell, Donald T. (2002). *Experimental and quasi-experimental designs for generalized causal inference*. Boston: Houghton-Mifflin. An update of a classic by a third author.
19. Bahaman, A. S. & Turiman, S. (1999). *Statistic for social research with computer application* (is'ed). University Putra Malaysia.
20. Fraenkel, J. K., & Wallen, N. E. (1993). *How to design and evaluate research in education* (2nd ed.). New York: McGraw-Hill.
21. Bordens, K. S., & Abbott, B. B. (1996). *Research design and methods: A process approach* (3rd ed.). MV, California: Mayfield Publishing Company.
22. Wen, Z. P., Zhou, X. Q., Feng, L., Jiang, J., & Liu, Y. (2009). Effect of dietary pantothenic acid supplement on growth, body composition and intestinal enzyme activities of juvenile Jian carp (Cyprinus carpio var. Jian). *Aquaculture Nutrition, 15,* 470–476. https://doi.org/10.1111/j.1365-2095.2008.00612.
23. Weinstein, A. R., Sesso, H. D., Lee, I. M., et al. (2004). The relationship of physical activity vs body mass index with type 2 diabetes in women. *JAMA, 292*(10), 1188–1194.
24. Flores, J. (1994). *Puerto Rican and proud boyee!: Rap roots and amnesia.*
25. Mastura, Rozita, & Sofian, Mohd. (2008). Qualitative study: Influence of aerobic exercise on state and trait anxiety among working women. *International Journal for Education Studies, 1,* 67–76.
26. Inelmen, E. M., Toffanello, E. D., Enzi, G., et al. (2005). Predictors of drop-out in overweight and obese outpatients. *International Journal of Obesity, 29*(1), 122–128.

Part II
Sports Technology and Management

Real-Time Soccer Team Monitoring for Indoor Training Using Wireless Local Area Network

N. Effiyana Ghazali, M. A. Baharudin and S. K. S. Yusof

Abstract Match congestion in elite soccer results in muscle fatigue and underperformance on the next coming game. Recently, researchers are proposing on monitoring soccer players using technology to overcome this problem. Existing monitoring systems that utilizes Wireless Local Area Network (WLAN) is lacking. Most of the current monitoring systems use Global Positioning System (GPS); GPS is not accurate to use for indoor purposes. Therefore, in this work, WLAN monitoring system is proposed to be used for indoor soccer training. Each player has a sensor and the sensor is connected wirelessly to the Access Point (AP) for real-time data transmission. Data transmission for four scenarios which are standing, walking, jogging and running will be investigated in terms of packet delay and packet loss.

Keywords Indoor · Monitoring system · Real-time data · Soccer · Training

1 Introduction

In Malaysia, contemporary elite soccer should be able to compete in up to 30 matches per season with an average of one to two match(es) a week [1]. This congested schedule for an elite team is mainly due to their involvement in both domestic as well as international leagues. This scenario increases the possibility of muscle fatigue and injuries, which will degrade the performance of the players for future games [2–4]. Thus, monitoring the player's condition is paramount in order to maintain the player's fitness, recovery state and performance whilst minimizing muscle fatigue and injuries.

Currently, there are many off-the-shelf (OTS) devices available for monitoring. Some of them such as Acentas team monitoring system, Activio Sport System, POLAR Team^2 Pro and Suunto Pro Team Pack have been studied in [5]. However,

N. Effiyana Ghazali (✉) · M. A. Baharudin · S. K. S. Yusof
Advanced Telecommunication Technology (ATT), Faculty of Engineering, School of Electrical Engineering, Universiti Teknologi Malaysia, Johor Bahru, Malaysia
e-mail: nurzal@utm.my

N. Sulaiman et al. (eds.), *Proceedings of the 3rd International Colloquium on Sports Science, Exercise, Engineering and Technology*,
https://doi.org/10.1007/978-981-10-6772-3_11

among the tested products, only one offers data transmission over WLAN. Their study includes the hardware and software design, measurement accuracy and price of the devices. However, there is no mention of the device performance on the data transmission performance over WLAN when the players are moving. This information is very important, especially, for real-time monitoring. Thus, the research in this paper considers this issue.

Besides that, most of the existing models and studies [4, 6, 7] only considers outdoor training which highly depends on location tracking via GPS. Nevertheless, GPS has poor accuracy when tracking subjects that are moving indoors. Thus, an alternative method is needed in order to resolve this issue. In summary, the contributions of this manuscript are analysis on the data transmission performance via WLAN for mobile players.

The rest of this paper is as follows: Sect. 2 discusses the most related works. In Sect. 3, the proposed method will be disclosed continued by the future works in Sect. 4 and the conclusion in Sect. 5.

2 Literature Review

Garcia et al. [6] proposed an approach where the nodes on the players become a WSN node themselves and the information collected are multi-hopped via the nodes of both teams in the field. This approach is able to minimize the load for each node. This approach can lessen the load on the nodes. However, the security and the reliability of the packets sent can be compromised and can be easily attacked.

Gaudino et al. [7] on the other hand proposed an improved method that reduces the WSN nodes power consumption by utilizing a fatigue-based threshold to trigger information transfer to indicate that the player is fatigued. This method can further reduce the transmission occurrence and implicitly reduce the nodes power consumption. However, using this approach data is transferred only when the players' fatigue level exceeds the threshold. This drastically reduces the obtainable information that can be used to detect the fatigue level in detail and can be used for future reference. Thus, research wise, this approach is unfavourable.

Furthermore, both methods only consider the communication for the outdoor scenario. Therefore, in this study, a new approach that can resolve the existing issues with consideration on the indoor scenario will be developed and discussed in the next section.

3 Proposed Method

In general view of our proposed method, a player is wirelessly connected to the Access Point (AP).Then, the AP is connected to the network and the fitness coach

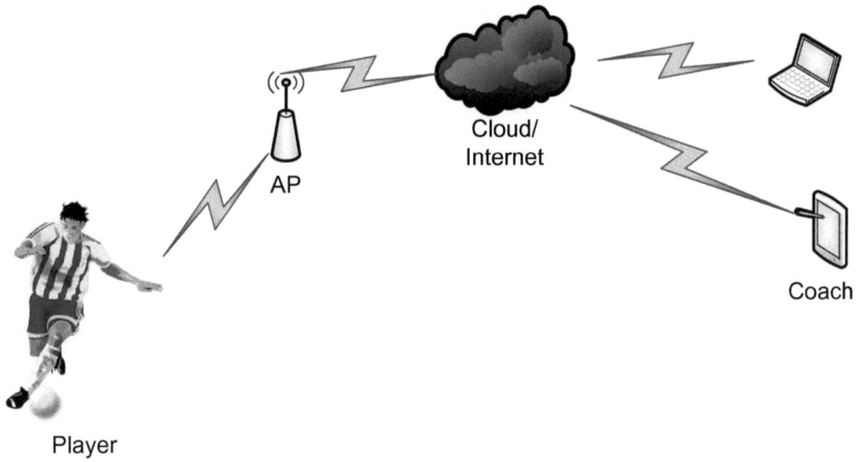

Fig. 1 The overall system architecture

can receive and view the data sent by the sensor in real-time using their smartphone or laptop. The general architecture is shown in Fig. 1.

In this experiment, data transmission for four scenarios will be investigated which are when the players are standing, walking, jogging and running. The speed between 0 and 0.7 km/h is considered as standing, speed between 0.7 and 7.2 km/h is considered as walking, speed between 7.2 and 14.4 km/h is considered as jogging and speed between 14.4 and 19.8 km/h is considered as running [8]. The standing position is shown in Fig. 2. The size of the field is followed as in [9].

As can be seen in Fig. 2, there is one AP located 20 m away from the field and six Relay Nodes (RNs) is tabulated near the field. RN is used to receive the data from the sensor and send to the AP before AP can send the data to the cloud. This is because the sensor does not have enough power to send directly to the AP. The detailed architecture is shown in Fig. 3.

The player mobility pattern (walking, jogging and running scenario) for experimental purposes is shown in Fig. 4, where the players will move in a straight line.

4 Future Work

The scenarios will be simulated using OMNeT++. From the simulation, data transmission will be analyzed based on packet delay and packet loss for four different scenarios as discussed before. After that, the real experiment will be conducted in indoor stadium and the results will be compared.

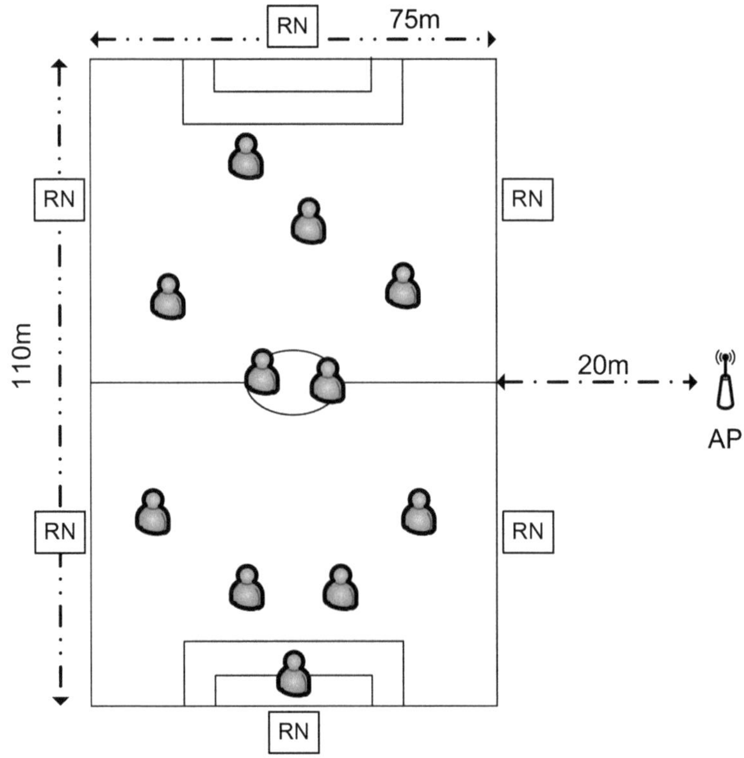

Fig. 2 Position of the players

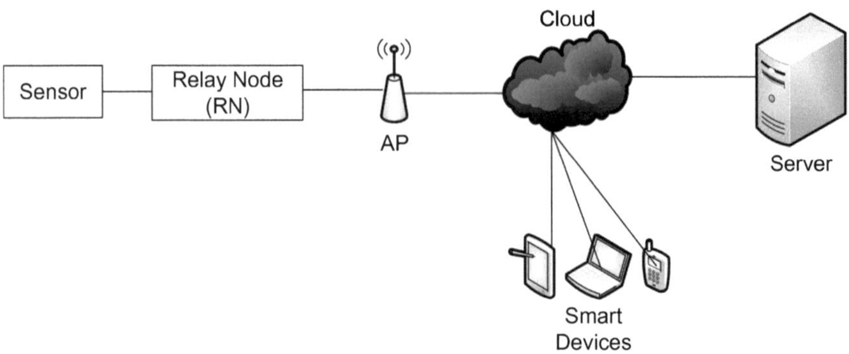

Fig. 3 Detailed architecture

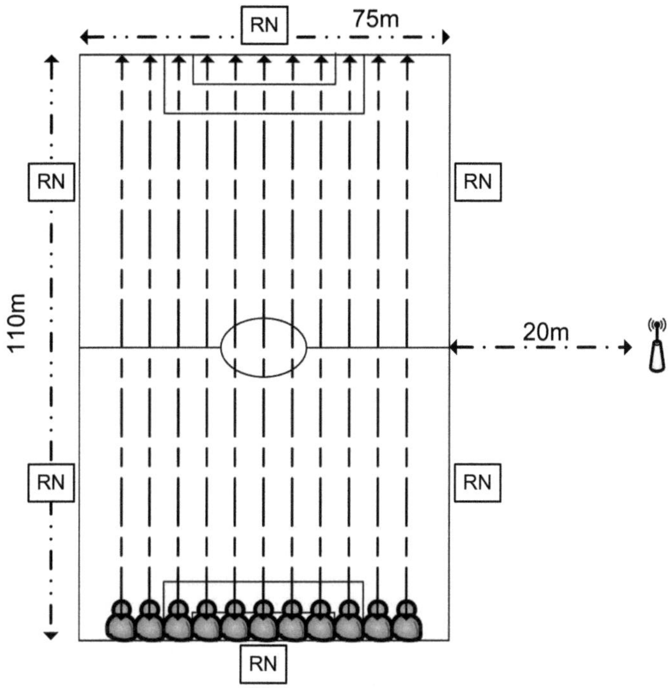

Fig. 4 Walking, jogging and running scenarios

5 Conclusion

In this paper, we present a WLAN indoor monitoring system to support the soccer team that wants to have indoor training during bad weather. In our system, the sensor attached to the player is connected wirelessly. The data sent from the sensors are relayed to the AP before it is sent to the cloud. Coach or fitness coach will be able to view the data in real time. The data transmission will be investigated in four scenarios which are when the players are standing, walking, jogging and running. In the future, simulation will be done using OMNeT++ before conducting a real experiment.

Acknowledgements The authors would like to thank all who contributed toward making this research successful. The authors wish to express their gratitude to Ministry of Higher Education (MOHE), Research Management Center (RMC) for the sponsorship, and Advanced Telecommunication Technology Research Group, Universiti Teknologi Malaysia for the financial support and advice for this project. (Vot number Q.J130000.2723.02K47).

References

1. Johor Darul Ta'zim FC. *All JDT matches 2015* [Online]. Available: http://www.johorsouthernti gers.com.my/calendar/all-jdt-matches-2015/.
2. The New York Times. *Heart rate monitors fine-tune soccer players' fitness* [Online]. Available: http://www.nytimes.com/2011/11/19/sports/soccer/heart-rate-monitors-fine-tune-soccer-players-fitness.html.
3. Akram, S., Javaid, N., Tauriq, A., Rao, A. & Mohammad, S. N. (2010). The-fame: Threshold based energy-efficient fatigue measurement for wireless body area sensor networks using multiple sinks. In *Eighth International Conference on Broadband, Wireless Computing, Communication and Applications* (pp. 201–220).
4. Carling, C., Gregson, W., McCall, A., Moreira, A., Wong, D. P. & Bradley, P. S. (2015) *Match running performance during fixture congestion in elite soccer: Research issues and future directions* (pp. 605–613). Switzerland: Sports Medicine, Springer International Publishing.
5. Schonfelder, M., Hinterseher, G., Peter, P. & Spitzenpfeil P. (2011). Scientific comparison of different online heart rate monitoring systems. *International Journal of Telemedicine and Applications*, Hindawi Publishing Corporation.
6. Garcia, M., Catalá, A., Lloret, J. & Rodrigues, J. J. P. C. (2011). A wireless sensor network for soccer team monitoring. *International Conference on Distributed Computing in Sensor Systems and Workshops* (DCOSS).
7. Gaudino, P., Iaia, F. M., Alberti, G., Strudwick, A. J., Atkinson, G., & Gregson, W. (2013). Monitoring training in elite soccer players: Systematic bias between running speed and metabolic power data. *International Journal of Sports Medicine, 34*(11), 963–968.
8. Rampinini, E., Bishop, D., Marcora, S. M., Ferrari Bravo, D., Sassi, R. & Impellizzeri, F. M. (2007). Validity of simple field tests as indicators of match-related physical performance in top-level professional soccer players. *International Journal of Sports Medicine, 28*(3).
9. FIFA. *Laws of the game 2015/2016*.

The Differences in Selected Performance Indicators Among Top Four and Bottom Four Teams in MASUM Rugby Sevens Tournament

Ahmad Naim Ismail, Siti Hartini Azmi and Norasrudin Sulaiman

Abstract This study aims to investigate the differences in selected performance indicators among the top four and bottom four teams who participate in MASUM 2015 Rugby 7s tournament. The data was collected from eight teams, namely the top four and bottom four teams, which are competed in the stages of grouping, quarterfinal, semifinal, and final, for this tournament. A total of 15 performance indicators were selected as the variables for this study. Descriptive and inferential statistics were employed to analyze the data. The results show that there exists a significant difference among the top four and bottom four teams for the performance indicators of (tackles missed, $p = 0.02$; conversion kick in, $p = 0.02$; conversion kick out, $p = 0.02$; try, $p = 0.03$; and maul, $p = 0.03$). The others performance indicators in this study show no significant difference among the top four and bottom four teams. The results presented in this study might be useful as a reference for coaches in designing training programs for the next MASUM Rugby Sevens tournament.

Keywords Performance analysis · Performance indicators · Rugby 7's
Top-bottom team characteristics

1 Introduction

Performance analysis is widely applied in sports, especially for team sports, to analyze the performance and weaknesses of the team itself, as well as the opposing team. Performance analysis can be carried out by manually observing a live match, with notational analysis, or by using computerized analysis with video as support. A video is the best evidence for performing the analysis, which can convince the team

A. N. Ismail · N. Sulaiman (✉)
Faculty of Sports Science and Recreation, Universiti Teknologi MARA, 40450 Shah Alam, Selangor, Malaysia
e-mail: noras878@salam.uitm.edu.my

S. H. Azmi
Faculty of Coaching Science, Universiti Pendidikan Sultan Idris, Tanjung Malim, Malaysia

N. Sulaiman et al. (eds.), *Proceedings of the 3rd International Colloquium on Sports Science, Exercise, Engineering and Technology*,
https://doi.org/10.1007/978-981-10-6772-3_12

about their performance [1]. Rugby union is among the most popular international sports, characterized by a high frequency of physical contact and repeated intermittent bouts of high-intensity activity. To perform under these physiological demands, a development combination of fitness of endurance, power, speed and acceleration, and sport-specific skills, are needed [2].

However, the specific demands of competitions differ markedly between 15 sides' rugby union and rugby sevens. Rugby sevens is a format of rugby union that has increased in popularity in recent years, and will be contested at the 2016 Olympics. The laws of rugby sevens, including the field dimensions, are substantially the same as 15 sides' rugby union, with the major exceptions being the reduced number of players (seven players per team), and shorter match duration (two halves of 7 minutes).

Differences in the physiological requirements of the two rugby formats suggest that the characteristics of high-level players in each format may also differ [2]. The dynamic match environment can make it difficult for coaches and support staff to identify which elements of physical, technical, and tactical development to focus on in order to enhance the probability of successful performance. Match analysis is often used to provide an unbiased record and objective of team activity to assess and monitor overall performance. However, it is unclear which performance indicators should be monitored to evaluate the team performance in rugby sevens [3]. A performance indicator is a variable characteristic for some aspect that is related to successful performance outcomes [3]. A research study is required to characterize the technical and tactical aspects of team plays, as related to successful performance in rugby sevens.

Identifying the selected performance indicators related to scoring points and winning in rugby sevens is beneficial to develop reference values for international matches. These values can be applied by coaches to inform practical guidelines for optimal technical and tactical development. The reference values can assist in understanding the variability of team performance and guide the coaches to establish quantifiable objectives for training and competition performance, and help to evaluate the efficacy of training interventions and tactical changes. The knowledge of performance indicators can also be applied to create a performance profile in order to predict team behavior and performance outcomes in the future [3].

In a rugby match, scoring a point by try and conversion is the most important manner to determine the winner. There are many performance indicators in rugby, namely, kickoff, ruck win, ruck lose, scrum win, scrum lose, tackle made, tackle missed, conversion kick win, conversion kick out, dropkick, try, penalty, lineout win, lineout lose, turnover, and maul. From the performance indicators mentioned above, this does not mean that the top four teams would have a higher frequency of performance indicators compared to the bottom four teams, in order to win and gain higher rankings. The top four teams might have a lower frequency of performance indicators compared to the bottom four teams. For example, the winning team might have strength in tackling the opponent so that they would require less tackles to be made in order to turn over the ball. On the other hand, the losing team might lack strength in tackling the opponent, so that they would require more tackles to be made in order to turn over the ball. However, the winning team would require more try and conversions compared to the losing team in order to win, and the losing team surely has less try and conversions compared to the winning team. So, this means that the top four teams would not have a higher frequency of performance indicators compared to the bottom four teams. The performance indicators become important for discriminating winning and losing, but they might be changeable from time to time. This study aims to investigate which of selected performance indicators affect high influence for the top four and bottom four teams in MASUM 2015 Rugby Sevens Competition.

The purpose of this research is to investigate the differences in selected performance indicators among top four and bottom four teams in MASUM 2015 Rugby Sevens tournament.

2 Method

2.1 Performance Indicator

The performance indicators used in this study are the variables that aim to define some aspect of performance of sports, and they are related to successful performance or outcome. The selected performance indicators are shown in Table 1.

Table 1 Selected performance indicators

No.	Variables
1	Kickoff
2	Ruck win
3	Ruck lose
4	Scrum win
5	Scrum lose
6	Tackle made
7	Tackle missed
8	Conversion kick in
9	Conversion kick out
10	Dropkick
11	Try
12	Penalty
13	Lineout win
14	Lineout loss
15	Maul

Source Adopted from national rugby coach and journal of difference of game statistics between winning and losing teams in 2011 rugby world cup

Table 2 Top four and bottom four teams

Top four teams	Bottom four teams
UPM A	UMT
UPM B	UUM
UiTM A	UTeM
UTM	UM A

2.2 Sample Selection

The sampling technique was conducted at "Padang Ragbi A" and "Padang Ragbi B", Universiti Putra Malaysia (UPM), Serdang. This tournament consists of 32 teams, within 8 groups. The format of this tournament is that there will be a grouping stages, and are divided into Cup, Bowl, Plate, and Shield pools. Each team will compete in 6–7 matches for this tournament. There will be a total of 52 matches for the top four and bottom four teams. The top four and bottom four teams are determined by the final results of which team are in the Cup and Shield pools. Four teams which compete in the semifinals Cup pool will be categorized as the top four teams, and the other four teams which compete in the semifinals Shield pool will be categorized as the bottom four teams. Table 2 shows the selected top and bottom four teams for this study.

2.3 Data Analysis

The data collected from this research were gathered and interpreted using the SPSS statistics v.21 software in order to determine the descriptive statistics by calculating the mean, $\overline{x}$ and standard deviation, and s, which significant value is set at $p < 0.05$.

For the inferential statistics, the nonparametric statistical test, which is Wilcox on Matched Pairs Signed Rank Sum Test, was used to compare the differences in performance indicators among top four and bottom four teams. This is to determine whether a significant difference exists between all groups.

2.4 Reliability Procedures

Intra- and Inter-Reliability test and Total Error calculation were performed in order to determine the reliability of the analysis performed in this study. The reliability for the intra-observer is $r = 0.89$, which shows a very good association. The reliability for the interobserver is $r = 0.80$, which shows a correlation between observer. The percentage of total error for the reliability procedure was 7%.

3 Results

3.1 Mean and Standard Deviation

Table 3 shows the mean and standard deviation obtained in this study.

3.2 Differences Between Top Four and Bottom Four Teams

The Mann–Whitney U Test was performed to investigate the differences among the top four and bottom four teams. The results are shown in Table 4.

4 Discussion

The first result shows that there are significant differences among top four and bottom four teams for tackle missed ($Z = -2.323$, $p = 0.020$, $p < 0.05$), conversion kick in ($Z = -2.323$, $p = 0.020$, $p < 0.05$), conversion kick out ($Z = -2.323$, $p = 0.020$, $p < 0.05$), try ($Z = -2.191$, $p = 0.028$, $p < 0.05$), and

Table 3 Mean and standard deviation

No.	Indicator	Top four teams		Bottom four teams	
		Mean	SD	Mean	SD
1	Kickoff	4.18	2.57	2.64	1.59
2	Ruck win	1.43	2.27	2.27	2.73
3	Ruck lose	0.32	0.55	0.39	0.73
4	Scrum win	1.18	0.82	0.89	1.17
5	Scrum lose	0.18	0.39	0.25	0.65
6	Tackle made	10.46	5.11	10.11	5.09
7	Tackle missed	1.85	1.81	2.93	2.48
8	Conversion kick in	2.93	2.29	1.39	1.40
9	Conversion kick out	1.50	1.25	0.86	0.93
10	Dropkick	0.29	0.85	0.39	0.74
11	Try	4.46	2.60	2.29	1.80
12	Penalty	1.04	1.17	1.64	1.57
13	Lineout win	0.39	0.57	0.25	0.52
14	Lineout lose	0.21	0.41	0.32	0.55
15	Maul	0.07	0.262	0.25	0.585

Table 4 The Mann–Whitney U Test results

No.	Indicators	Z value	PI value
1	Kickoff	−2.19	0.28
2	Ruck win	−1.44	0.15
3	Ruck lose	−5.92	0.55
4	Scrum win	−1.16	0.25
5	Scrum lose	0.00	1.00
6	Tackle made	0.00	1.00
7	Tackle missed	−2.32	0.02*
8	Conversion kick in	−2.32	0.02*
9	Conversion kick out	−2.32	0.02*
10	Dropkick	−4.58	0.65
11	Try	−2.19	0.03*
12	Penalty	−1.89	0.06
13	Lineout win	−1.35	0.18
14	Lineout lose	0.00	1.00
15	Maul	−2.14	0.03*

*Show significant difference, $p < 0.05$

maul ($Z = -2.139$, $p = 0.032$, $p < 0.05$). However, there were no significant differences among top four and bottom four teams for kickoff ($Z = -2.191$, $p = 0.28$, $p > 0.05$), ruck win ($Z = 1.443$, $p = 0.149$, $p > 0.05$), ruck lose ($Z = -5.92$, $p = 0.554$, $p > 0.05$), scrum win ($Z = -1.162$, $p = 0.245$, $p > 0.05$), scrum lose ($Z = 0.000$, $p = 1.000$, $p > 0.05$), tackle made ($Z = 0.000$, $p = 1.000$, $p > 0.05$), dropkick ($Z = -4.58$, $p = 0.647$, $p > 0.05$), penalty ($Z = -1.888$, $p = 0.059$, $p > 0.05$), lineout win ($Z = -1.348$, $p = 0.178$, $p > 0.05$), and lineout lose ($Z = 0.000$, $p = 1.000$, $p > 0.05$).

Based on the result shown, we can conclude that tackle missed, conversion kick in, conversion kick out, try, and maul have a significant difference among top four and bottom four teams. Other performance indicators in this study showed no significant difference among top four and bottom four teams.

This study also found that the top four teams had a higher mean for kickoff; this is because many tries were made. This shows that the top four teams had more chances to score more compared to the bottom four teams. Scrum win indicators also show that the top four teams had a higher mean compared to the bottom four teams. In addition, the tackle made also shows a higher mean compared to the bottom four teams. This probably shows that the top four teams are stronger in terms of muscle strength and technics during scrum and tackle. The performance indicators for try, conversion kick in, and conversion kick out also show a higher mean compared to the bottom four teams. This clearly shows that the top four had more tries so that they had more chances in conversion kick. Another performance indicator that showed a higher mean for the top four teams was the lineout win.

The bottom four teams had a higher mean for ruck win, ruck lose, scrum lose, tackle missed, dropkick, penalty, lineout lose and maul. This shows that the bottom four performed more rucking to gain possession of the ball after being tackled by the opponent. The scrum lose and tackle missed also shows that the bottom four teams probably lacked strength for scrums and tackle. Many scrums that they engaged were lose and many tackle that they performed were missed. The lineout lose shows that the bottom four teams may lack technicals and tacticals compared to the top four teams.

The maul performance indicators show that the bottom four teams performed maul to increase the strength to avoid them from being tackled by the opponent.

5 Conclusion

In conclusion, the top four teams achieved a successful kickoff, try, scrum win, tackle made, conversion kick, and lineout win compared to the bottom four teams. These results can be used as a reference by coaches to develop a proper training program based on scrum and tackle, and monitoring the efficiency during difficult situations in competitions. The defensive training can be improved to the specific training to enhance defense techniques. This study provides important indicator values that may serve as a reference to design and evaluate training for team peak performance in competition tournaments. Coaches can use the results to establish suitable objectives for their players and teams by maintaining strength and addressing the weakness of the team.

References

1. Polidoro, L., Bianchi, F., Di Tore, P., Alfredo, & Raiola, G. (2013). Futsal training by video analysis. *Journal of Human Sport & Exercise, 8*(2), S290–S296.
2. Higham, D., Pyne, D., Anson, J., & Eddy, A. (2013). Physiological, anthropometric and performance characteristics of rugby sevens players. *Journal of Sports Physiology and performance, 8*, 27.
3. Higham, D., Hopkins, W., Pyne, D., & Anson, J. (2014). Performance indicators related to points scoring and winning in international rugby sevens. *Journal of Sports Science and Medicine, 13*, 364.

An Overview of Local Authority and Stadium Corporation Sports Facility Maintenance Management Practices in Malaysia

Hasnul Faizal Hushin Amri, Siti Aishah Wahab, Norlena Salamuddin and Mohd Taib Harun

Abstract This study aimed to review the public sports facilities maintenance management practice managed by the Local Authority (LA) and the Stadium Corporation (SC) in Malaysia. This study is also intended to explore differences in the practice of the public sports facilities maintenance management based on respondents and facilities background as well as comparing two models of public sports facility management organizations. The study is based on the maintenance effectiveness model developed by Cholasuke, Ramnik Bhardwa and Jiju Antony (2004). The sample consists of 105 maintenance management staffs from 13 LA and 7 SC in Malaysia. The purposive sampling technique was used for sample selection of this study. A Likert scale questionnaire was adapted and used for the study and after that, it underwent a process of validity and reliability within the local context. Descriptive statistics are used to describe the basic features of the data in a study by using frequencies, percentages, mean, and standard deviation. Meanwhile, nonparametric test such as Mann–Whitney U, Kruskal–Wallis, and Wilcoxon T-test were used for inferential analysis. From the data analysis, it showed that the LA is more effective than the SC in terms of maintenance management for public sports facilities as a whole. The result for each effectiveness maintenance success factors showed that LA is more effective in the management of public sports facilities maintenance compared to the SC in Malaysia based on the factors such as organizational structure maintenance, planning and scheduling maintenance, maintenance information management, human resource management, and maintenance contract management. However, SC is only effective in continuous improvement factor. This study is expected to provide important information to help sports administrators understand related issues in the maintenance management of sports facilities, thus helping to improve the quality of sports facility maintenance management in Malaysia.

Keywords Sports facility · Maintenance management · Maintenance

---

H. F. H. Amri (✉) · S. A. Wahab
Faculty of Sports Science and Recreation, Universiti Teknologi MARA, Shah Alam, Malaysia
e-mail: hasnulfaizal@salam.uitm.edu.my

N. Salamuddin · M. T. Harun
Department of Sports Studies, Universiti Kebangsaan Malaysia, Bangi, Malaysia

N. Sulaiman et al. (eds.), *Proceedings of the 3rd International Colloquium on Sports Science, Exercise, Engineering and Technology*,
https://doi.org/10.1007/978-981-10-6772-3_13

1 Introduction

The role of sports facilities is not only for sports development aspects but also important for local development aspects such as municipal, economic and social. Based on a report by Zulfakar [1] in The Star, there are 8261 sports facilities were built in Malaysia. From this, 619 sports facilities are managed and operated by the Ministry of Youth and Sports while the remaining 7642 are under the management of the state government and Local Authorities (LA).

In every Malaysian Plan (MP) from the years 1966 to 2017, government has built various sports facilities ranging from multipurpose complexes to the construction of stadiums in every state. However, the sports facilities provided are not fully utilized, misused, unprotected, and not maintained by the responsible parties. The issues of weaknesses maintaining public sports facilities in Malaysia are often debated in parliament and media because it is not in line with the objectives to be achieved.

At one point, the government had taken drastic action to freeze construction of new sports facilities due to the lack of maintenance by various responsible agencies. This is based on the excerpt from Dato' Sri Najib Tun Razak's statement, Deputy Prime Minister of Malaysia stated that sports facilities throughout the country are not well maintained and most are in unsatisfactory condition [2].

Ministry of Youth and Sports Malaysia is the main agency responsible for the development of sports in Malaysia which also has seen limited authority to intervene in the management of sports facilities managed by the state government through LA in Malaysia [3]. This situation has caused many sports facilities in certain state and district throughout Malaysia were not in satisfactory conditions due to the lack of maintenance activities. The issues on weaknesses maintaining public sports facilities could be categorized as a national issue which is often exposed by various sources through the mass media as well as report from the national auditor. Most of the reports show sports facilities are facing various problems in terms of maintenance management due to the weakness of existing operating systems. The financial aspect is also seen as the main difficulties in carrying out effective and quality maintenance activities by the agency such as LA and SC. From the operation view, the LA in each state is sufficiently burdened with the obligation to take care of the public sports facilities.

The purpose of this research is to identify the factors that lead to a successful maintenance management across Malaysia public sports facility manage by the LA and SC in selected state. The results of the study draw an assessment of strengths and opportunities that could provide a useful indicator for supporting evidence for decision makers to improve maintenance management practices and formulate guidelines for future best practice on maintenance management sports facilities in Malaysia.

2 Method

A survey was conducted to describe and explain the maintenance management effectiveness of public sports facilities in Malaysia. Maintenance Management Effectiveness (MME) questionnaire developed by Cholasuke et al. [4] was used in this study. The study focused towards main key elements related to maintenance management effectiveness that has been verified by practitioners and professionals in the field of maintenance. Nine (9) major elements have been identified as important in maintenance management effectiveness factor such as (1) policy deployment and organization, (2) maintenance approach, (3) task planning and scheduling tasks, (4) information management and *CMMS*, (5) spare parts management, (6) human resource management, (7) contracting out maintenance (8) financial aspect, and (9) continuous improvement.

MME questionnaires were distributed to thirteen (13) LA and seven (7) SC major states in Malaysia. Cover letters and questionnaires were addressed directly via postal to all related respondents. Prior to this, a pilot study has been carried out to see the feasibility and acceptance of important inputs for this survey. Data received from the pilot study was analyzed and the 36 items showed a reliability factor of 0.833.

The sample selection for this study is based on purposive sampling method where the selection of subjects from the population consists and focused specifically on the thirteen (13) LA and 7 SC agencies, involving 105 respondents. The justification for selection of this population is based on the methodology and operation of sports facilities management in Malaysia which is either under the responsibility of the LA and SC. Those involved as respondents were directly responsible for their respective sports facilities maintenance.

3 Results

Further inference analysis was undertaken to address the question, whether there were significant differences in elements of maintenance management practice elements practiced by LA and SC. To test on research hypotheses, the Mann–Whitney U test by using the mean score for each factor was constructed. For this analysis, there are nine main hypotheses tested to see the difference in aspects of maintenance management practices between the two agencies.

The data analysis found that there was a significant difference between the Policy and Organizational Structure of maintenance organizations between LA and the SC in Malaysia. The results of the study were significant, where $U = 587.000$, $p < 0.05$. The null hypothesis was rejected and the researcher made a decision that there was a difference in the practice of Policy Creation and Restructuring the maintenance organization for both agencies. It was also found that LA (49.97) was more effective in term of Policy and Organizational Structure of the maintenance organization compared to the SC (36.08).

Based on the analysis, it was found that there was no significant difference between the Maintenance Approach between LA and SC in Malaysia, statistics U = 1083.500, p > 0.05. This means that researchers failed to reject the null hypothesis. Mann–Whitney U test results show that both agencies are ineffective in the overall aspect of the Maintenance Approach practiced in the operation of sports facilities maintenance in Malaysia.

The data analysis revealed that there were significant differences in Planning Factors and Maintenance Work Schedules between LA and SC in Malaysia. The results of the study were significant U = 758.500, $p < 0.05$. The null hypothesis was rejected and the researcher decided that there was a difference in Planning and Maintenance Work Schedule for both agencies. It was found that LA (57.52) was more effective in Planning and Maintenance Work Schedule compared to SC (41.16) in Malaysia based on differences in mean scores and standard deviation for both agencies.

The data analysis found that there was significant difference between Maintenance Information Management between LA and SC in Malaysia. The results of the study were statistically significant U = 859, $p < 0.05$. The hypothesis was rejected and the researchers decided that there was a difference in Maintenance Information Management for both agencies. It was also found that LA (56.20) was more effective in Maintenance Information Management than SC (44.62) in Malaysia.

Based on the analysis, there was no significant difference between Spare Parts Management and Maintenance Materials factors between LA and SC in terms of the effectiveness of maintenance management of public sports facilities, statistics U = 1080.5, p > 0.05. Therefore, the researcher fails to reject the null hypothesis, the results show that there is no significant difference in Spare Parts Management and Maintenance Materials for both agencies.

Analysis of Human Resource Management factors found that there were significant differences in Human Resource Management factors between LA and SC in Malaysia in terms of the effectiveness of maintenance management of public sports facilities, U = 752, $p < 0.05$. This analysis also shows that both agencies differ in their aspects of the practice of their Human Resource Management factors. This means the null hypothesis is rejected. The results of the analysis show that LA is more effective in the practice of Human Resource Management factor than SC.

The Maintenance Contract Management factors analysis shows that there is a significant difference between the LA and the SC in Malaysia in terms of the effectiveness of the maintenance management of public sports facilities, U = 846, $p < 0.05$. This analysis also shows that the two agencies differ from the aspects of their practice to the organizers of the contract management. This means the null hypothesis is rejected. Mann–Whitney U test results show that LA (56.37) is more effective in the practice of Maintenance Contract Management factor than SC (44.17).

Based on the analysis, it was found that there was no significant difference between LA and SC in Maintenance Financial factors, where U = 879.5, p > 0.05. This means that researchers failed to reject the null hypothesis. Mann–Whitney U test result shows that both agencies are ineffective as a whole in aspects of Maintenance Financial practiced in the operation of sports facilities maintenance in Malaysia.

An analysis of the Continuous Improvement factors of facilities maintenance found that there was a significant difference between the LA and SC in Malaysia in terms of the effectiveness of the maintenance management of public sports facilities, statistics $U = 576$, $p < 0.05$. This analysis also shows that both agencies differ from their practice and approach to the Continuous Improvement factor. This means the null hypothesis is rejected. Test results show that the SC (52.35) is more effective in practicing Continuous Improvement than LA (40.44).

4 Discussion

According to [4], the establishment of policy and organizational structure factor is an important aspect to be taken care. It is because well-planned establishment of policy and organizational structure are capable to mobilize manpower, develop effective communication, and create good maintenance resource allocation with the ability to ensure all planned maintenance activities are implemented effectively.

In the view of LA establishment policies and structures in managing and maintaining public sports facilities, in terms of staff number, LA has a large number of engineers and technical staff, experience, and resources compared to SC. However, the organizational structure of the SC is more focused, and the organization's policy and organizational structure were planned in line with the capacity and workload. These findings support the research conducted by Mukelas et al. [5] where a review of 12 LAs in Selangor shows that all LAs have an organizational structure and maintenance strategy that helps towards the effectiveness of complaint resolution and maintenance-related repairs.

The maintenance approach factors practiced by both agencies are seen based on the results of the analysis do not indicate any significant differences in assisting with the effectiveness of maintenance. Although based on the analysis found no meaningful relationship but the two agencies practiced an appropriate approach to existing capabilities and resources for both agencies. Based on the management practice study that focuses on effective maintenance services, it is still not at the best practice level where independent agencies determine the best and profitable way of managing agencies [6]. Statement by the chief secretary to the government, Hassan (2007) although the government's commitment to support effective maintenance through a substantial provision for upgrading and renovation work, there are still of work being carried out poorly manage and does not reach the required quality. According to Annies [7], Chief Maintenance Sector, Public Works Department Malaysia (JKR) shared the same view that maintenance standards in government-owned buildings do not meet the required quality.

The task planning factor and maintenance work schedule are one of the factors related to the planning and scheduling of maintenance work based on the maintenance policies that have been developed by the organization. The analysis found that there were significant differences in the planning and scheduling aspects of maintenance work between the two agencies. Based on the analysis, the LA was found to be more effective in this aspect than the SC. This factor is due to clear organizational structure and policies in the planning of maintenance practiced by the local LAs. Fast work execution process based on the amount of workforce, financial capabilities, and a combination of maintenance approaches that give advantages to LA compared to SC. Although SC is more specific in its direction, due to the lack of technical personnel and experienced in planning and scheduling work has inhibited the effectiveness of maintenance management. It is an evident from the auditor general's report stating that SC Malaysia did not plan effective maintenance activities which caused the roof of the stadium to be damaged due to the failure of planning and periodic maintenance work [8].

The maintenance management information factor found that LA was more effective where there were significant differences between LA and SC. Based on the findings, the LA is more organized in the maintenance information where there are a computer usage and storage of information used for the purpose of storing all maintenance records such as CMMS. The usage is not so comprehensive but efforts to improve the storage of these records are being upgraded. While the SC agency has not been practiced thoroughly and most of the information is stored manually referred based on the details of the work at sports facilities. Based on a study conducted by 24 companies and organizations with a department of facility management unit that used of computerized systems for building facilities management (Computer-Aided Facility Management) found that this application is seen to increase the efficiency of maintenance operation but the costs prevent optimum and comprehensive used in organizations in Malaysia [6]. Based on this study, the researcher also believes that the weakness and lack of application of this app which is recognized can save the time of operation and cannot be practiced by LA and SC because it involves high cost of computer software and hardware. It is not yet able adapting its users thoroughly due to financial constraints. The importance of using this software in the facility maintenance management should be noted [9], through the improvement and use of the software in the organization could help improve the quality and good financial returns. However, operation management tools based on analysis show that there is no significant difference shown through analysis for both agencies.

Both LA and SC rely heavily on appointed maintenance contractors for all spare parts replacement work. Generally, the importance of storing and having spare parts is closely related to the effectiveness of the maintenance and operational performance of a facility. The spare parts stock should be listed according to the most important and less important priority, this is to offset the costs and needs of an existing facility [10]. This aspect has a close relationship with the maintenance approach factor where the chosen approach will determine the spare parts storage and maintenance materials requirements, if an agency uses a comprehensive contractual approach then the need to have and store spare parts on a large scale is not necessary.

Human resources are an important factor in achieving the objectives of each organization. In this study, there were significant differences between LA and SC in human resource management. LA is more effective than SC in terms of human resource management that relies on training planning, staffing, competency level assessment, scope setting and task description in balancing staffing and workload.

LAs are seen to have good training planning where there is LAs that have implemented relevant training for staff in the 50–90% level related to their duties compared to SC. The level of competence and performance of LA staff is also better than SC staff where good staffing planning balances workloads and increases the motivation of LA staff.

Maintenance contract management is a factor to ensure the effectiveness of maintenance management practices of each organization. In this study, there was a significant difference between LA and SC where LA was more effective than SC in terms of maintenance contract management. LA through research findings were 30–70% maintenance work has been handed over to qualified contractors and this contributes to moderate high savings by most LA respondents. Most of these maintenance contracts provide a great deal of maintenance work that involves major repair work and special expertise.

The financial factors for both agencies based on the analysis show that there is no significant difference shown through analysis. This decision is because both of these LA and SC adopt a same approach and do not show the level of promising effectiveness in the management of finance and budget planning for maintenance activities. In the case of maintenance budget planning method, both respondents from the LA and SC practice rough estimates of planning and details based on maintenance activity requirements. In view of the expenditure aspect, SC spending more than 30% of what is allocated in the annual allocation versus the LA, with most respondents claiming that expenditure is only between 5 and 30%.

In the aspect of continuous improvement, factors for both LA and SC based on analysis showed a significant difference. The findings show that the SC is only effective in the aspect of continuous improvement because it is an organization that is focused on the management of the facility, and definitely the effort to improve in management and maintenance work can be done periodically and regularly. Compared to LA, the main task is not just to the sports facilities but covers the scope of maintenance that involves the municipality and the wider context to make it difficult for LA to do so frequently. Continuous improvement is mostly focused on policy revisions, conducting measurements of maintenance efficiencies, and maintenance management improvements through regular and periodic management reviews and reviews.

5 Conclusion

Based on the elements of the effectiveness of the maintenance management practice, it is found that the LA agencies are effective in five factors: (i) maintenance organization and organizational structure, (ii) planning and maintenance work schedules, (iii) maintenance information management, (iv) human resource management, and (v) maintenance contract management. Meanwhile, SC agencies are more effective only for continuous improvement factors.

References

1. Zulfakar, M. (2005, July 1). *Najib: Stop it. The Star*.
2. Abdullah, F. (2005). Projek Baru Kemudahan Sukan Dibeku. *Beirta Harian*, 1 Julai 2005.
3. Anon. (2009, June 3). *KBS Sedia Teliti Akta Sukan Untuk Kawal Selia Stadium*. Bernama.
4. Cholasuke, C., Bhardwa, R., & Antony, J. (2004). The status of maintenance management in UK manufacturing organisations: Results from a pilot survey. *Journal of Quality in Maintenance Engineering, 10*(1), 5–15. https://doi.org/10.1108/13552510410526820R.
5. Mukelas, F. M., Zawawi, E., Kamaruzzaman, S. N., Ithnin, Z, & Zulkarnain, S. (2012). *A review of critical success factors in building maintenance management of local authority in Malaysia* (PP. 653–657). https://doi.org/10.1109/isbeia.2012.6422970.
6. Kamaruzzaman, sn, & Zawawi, E. (2010). Development of facilities management in Malaysia. *Journal of Facilities Management, 8,* 75–81. https://doi.org/10.1108/14725961011019094.
7. Annies, A. (2007). current issues and challenges in managing government s assets and facilities. In *Proceeding of the National Asset and Facilities Management (NAFAM) Convention*, Kuala Lumpur, Malaysia, August 13.
8. Jabatan Audit Negara Malaysia. (2010). *Laporan Ketua Audit Negara Tahun 2010: Kewangan, Pengurusan Aktiviti, Dan Pengurusan, Serta Subsidiari, Syarikat Persekutuan, Badan Berkanun*.
9. Gallardo-Guerrero, L., García-Tascón, M., & Burillo, P. (2008). New sports management software: A needs analysis by a panel of Spanish experts. *International Journal of Information Management, 28,* 235–245. https://doi.org/10.1016/j.ijinfomgt.2007.09.005.
10. Yang, K., & Niu, X. (2009). *Research on the spare parts inventory. Facilities* (pp. 1018–1021).

Accelerometer Artefacts from Body-Worn Sensors

Elle McDonough, Christopher W. Hinton-Lewis, Hugo G. Espinosa and David V. Thiel

Abstract Inertial sensors are commonly found in commercial products such as smartphones and sports bands. When these sensors are stationary, the variation in the recorded signal (noise level) is very small (approaching 0.1% of the gravitational acceleration). In many sports applications, the recorded acceleration has noise spikes related to the sensor itself and impulsive body movements. The wrist acceleration (100 samples/s) from a boxing jab (posterior–anterior) to a stationary bag using five subjects demonstrated that following the impact, the sensor oscillated in a manner dependent on the sensor weight. Mathematically, the integrated acceleration at the end point of the jab should be zero. This was used to remove the vibration artefacts and verified using a high-speed camera (2014 fps). A male participant with a rotator cuff shoulder reconstruction was asked to raise his arms (front lateral raise) individually from rest (vertical) to above the horizontal position carrying several different weights. The shoulder instability of the participant was evident in the accelerometer record as significantly larger anterior–posterior vibration at 9 Hz. This was not evident in the opposite shoulder which had no reported injury. This technique might prove a useful tool in quantifying shoulder instability over long periods of time.

Keywords Inertial sensors · Accelerometers · Boxing jab · Shoulder stability Noise

E. McDonough · C. W. Hinton-Lewis · H. G. Espinosa · D. V. Thiel (✉)
Engineering, Griffith University, Nathan Campus, Brisbane, QLD 4111, Australia
e-mail: d.thiel@griffith.edu.au

E. McDonough · C. W. Hinton-Lewis · H. G. Espinosa
School of Engineering, Griffith University, Nathan Campus, Queensland, QLD 4111, Australia

E. McDonough · C. W. Hinton-Lewis
Sports and Exercise Science, Leeds University, Leeds, UK

N. Sulaiman et al. (eds.), *Proceedings of the 3rd International Colloquium on Sports Science, Exercise, Engineering and Technology*,
https://doi.org/10.1007/978-981-10-6772-3_14

1 Introduction

Inertial sensors are used widely in sport and everyday activities featuring in smartphones and sports bands; this can enable a variety of monitoring including heart rate and for this instance, acceleration. When these sensors are stationary, the variation in the recorded signal (noise level) is very small (approaching 0.1% of the gravitational acceleration). However, when increasing the intensity of the movement in sport and everyday living situations, the signals received are not noise free and often contain very large one-point spikes and other high-frequency vibration periods. External vibrations not produced by the movement can account for the noise; alternatively, movement of the sensors against the body due to loose coupling can cause artefacts [1]. This noise can cause issues when attempting to distinguish between the desired acceleration data from skeletal movement and error. Two likely causes are the inertia of the sensor, and joint instability during flexion. The method of attachment of the sensor, skin elasticity, muscle contraction effects and joint instabilities can cause these effects.

To remove this effect many researchers and embedded applications use various filtering techniques. They often rely on the assumption that movements fall in specific frequency bands, (walking typically 1–3 Hz range and running 5–7 Hz), whereas noise levels are located in higher bands [1]. However, filtering can result in the loss of high-speed biomechanical movements and filtering must degrade the signal of interest. For example, abrupt changes in acceleration of the hand in boxing and swimming could result in a set of inertia-induced vibrations in acceleration; abrupt changes in an ankle sensor could be the inertial response of foot-ground contacts. Such an effect was noted in arm rotation in classical ballet [2].

The specific locations of the sensors on the body can alter the noise given the magnitude of vibrations in particular areas. Earlier research suggests areas with lower quantities of soft tissue display lower levels of noise with accelerometers readings [1]. Knowing this enables coaches to locate and attach accelerometers to performers in areas which give more accurate readings. Numerous other investigations have seen a strong correlation between noise and exercise intensity. Participants with Osteoarthritis measured greater errors at greater walking speeds when analysing gait symmetry with accelerometers [3]. Given the consistency with these findings in 'normal gait', research has used these assumptions to generate equations and algorithms to account for the noise to create a clearer acceleration value which mimics that of actual skeletal movement [4]. In addition, more specific investigations have been made into accounting for skin motion artefact. Using EMG readings and vibrations as stimulants have enabled additional data to be collected suggesting greater accelerometer readings and oscillations at the sites of muscle contraction [5]. Skin motion artefacts are specific to the movement and the characteristics of the participant's skin (usually age related and location dependent) make this method far too time consuming and difficult [6]. Taking this into account, it would be more useful to investigate reducing the error causes and employing mathematical strategies to account for them. In addition, there might be useful biomechanical information in accelerometer measurements. This might be a significant advantage as dynamic joint movement measurements are recorded rather than static imaging—the current diagnosis tool.

This paper reports two separate experimental investigations using wrist-mounted sensors. The first is an acceleration measurement when punching a wall-mounted padded bag and the second reports shoulder vibrations in an arm-lifting exercise. These are commonly encountered movement artefacts in the acceleration records [2, 7, 9].

2 Experimental Methods

2.1 Boxing Jab

Hand speed is particularly important in boxing and can be measured using a wrist-mounted accelerometer [7]. Four male participants (1 experienced, 3 novices) were asked to jab a wall-mounted jabbing frame and leave the hand in the bag. A three-axis accelerometer [8] was mounted on the wrist using medical adhesive tape (see Fig. 1).

Synthetic data was created numerically for a 0.45 m long movement from a rest position to an abrupt termination in the bag. Figure 2 shows the change in position with time and the derived velocity and acceleration profile. The time to accelerate is much longer than the time to decelerate and so the positive acceleration peak has a much smaller magnitude than the deceleration phase. As the movement is from a resting position (0 m) to another resting position at 0.45 m, the mean value of the acceleration through the movement must be zero. However, the skewness of the acceleration data will not be zero as acceleration distribution is asymmetric. For the case shown, the skewness was 2.5.

Each jab was recorded using a high-speed video camera (1216 f/s) against a calibrated square grid backdrop located 20 cm behind the subject's forearm (Fig. 1). In one case, the accelerometer was loaded with additional 20 g weights to observe changes in the acceleration profile due to changes in the sensor inertia. The weights were taped to the acceleration sensor only, and not to the skin. The acceleration files were downloaded to a computer and processed using Matlab®. The jab speed

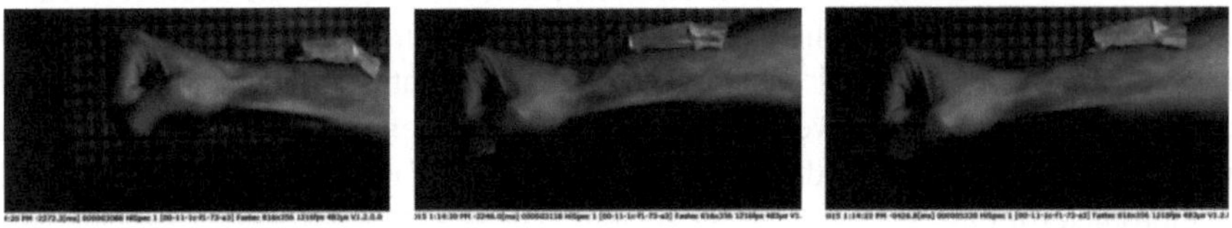

Fig. 1 Wrist-mounted sensor before impact (left); the sensor at time of impact (middle); the sensor orientation changes immediately following impact (right)

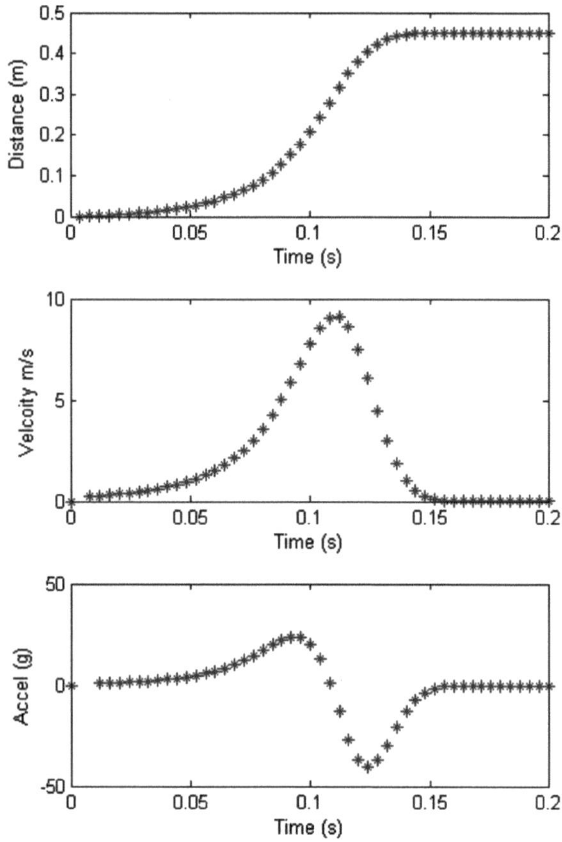

Fig. 2 Synthetic data showing the position of the arm in the direction of a jab as a function of time and the calculated velocity and acceleration. The difference in the change in velocity results in a smaller positive acceleration peak before impact and a much larger negative acceleration value after impact

was verified from the video frames. This work was conducted under the institution's ethics committee project approval (ethics approval number ENG/14/13/HREC).

The three-axis acceleration data was calibrated to readings in the gravitational acceleration 'g'. The principal axis of forearm movement (posterior–anterior) was used exclusively in data processing. Given that the total integrated acceleration over time must be zero, the first two changes (maximum and minimum) in acceleration measured as a deviation from zero were separated from the acceleration oscillations which followed. The latter is due to the movement of the sensor alone without forearm movement. For continuity, the deceleration maximum transition to zero was fitted by a third-order polynomial and the modified acceleration profile was integrated to determine the maximum velocity. The residual acceleration profile was analysed in the frequency domain to determine the frequency of oscillation of the weighted accelerometer unit.

2.2 Arm Lifting

One male participant was asked to raise hand-held weights (0, 2, 3 and 5 kg) with three repetitions, using the sequence shown in Table 1 (see Fig. 3). The left hand was subjected to a series of weights (0–5 kg). The participant (65 years) had a left shoulder reconstruction in 2012. All movements were recorded using wrist-mounted sensors on both arms and a video camera (25 f/s). Figure 3 shows the start position and the highest point in the lift. This is a common exercise used in gym training to strengthen the deltoid muscles in the shoulder.

The movement was sufficiently slow to allow static angle measurements to be made accurately. The experiment was conducted under the ethics committee project approval (ENG/14/13/HREC).

Table 1 Exercise sequence for arm-lifted weights

Sequence	Left weight (kg)	Right weight (kg)
1. Left-right-together (three repeats)	0	0
2. Left-right-together (three repeats)	2	0
3. Left-right-together (three repeats)	3	0
4. Left-right-together (three repeats)	5	0
5. Left-right-together (three repeats)	3	3

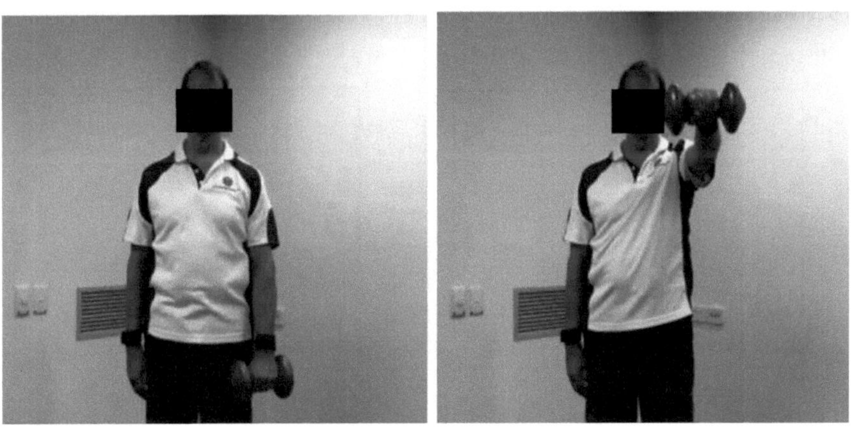

Fig. 3 Participant 1 at the resting position with a 3 kg weight in the left hand (left); tarticipant 1 at the highest point of the lift. The accelerometers are located in Velcro wristbands (right)

3 Results and Discussion

3.1 *Boxing Jab*

Figure 4 (left) shows the calibrated acceleration data for the same sensor and three different weights. Note that, the acceleration profiles are inverted when compared to Fig. 2.

Following the main positive acceleration peak, the acceleration oscillates before returning to zero. These oscillations are due to the inertial vibrations of the sensor relative to the hand and do not represent limb movement. While the maximum values are all overrange for the sensor (greater than 16 g), the damped oscillations which follow show increased amplitude as the weight was increased. The peak corresponding to 43 gr shows less deceleration which results in less velocity. This is represented by a lower amplitude in the oscillation, and it is due to the velocity not being well controlled. The peak oscillation frequency also changed from 10, to 17 and 19 Hz, respectively (Fig. 4—right) with increasing sensor weight. The acceleration time series was processed to remove the trailing oscillations (see Fig. 5) and the final result was integrated to calculate the velocity as a function of time.

3.2 *Arm Lifting*

The acceleration data from the arm-lifting sequence was converted to 'g' units (the earth's gravitational accelerat ion). When the arm is at rest, axis 1 (A1) is medio-lateral, axis 2 (A2) is anterior–posterior and axis 3 (A3) is vertical. The frequency components were derived from the time series data and plotted on a log–log scale.

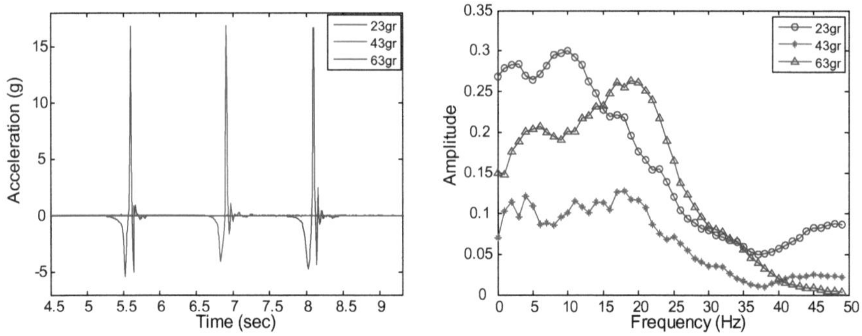

Fig. 4 The acceleration of a wrist-mounted sensor in the direction of the jab for three different weights. The three acceleration plots are superimposed for easy comparison (left). The amplitude of the oscillations following the major impact shows that the peak frequency increased with increasing weight (right)

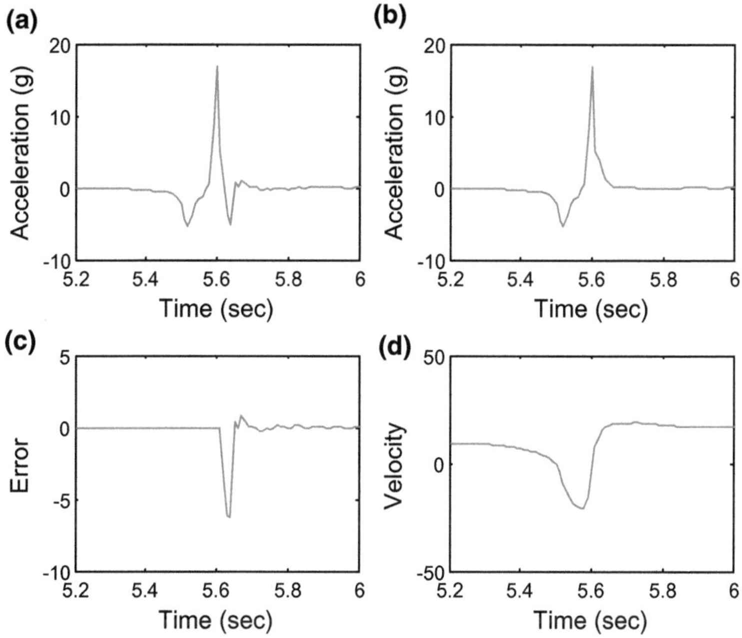

Fig. 5 Signal processing to remove the trailing oscillations **a** Unaltered acceleration data calibrated in g (top left), **b** Acceleration data with trailing oscillations removed (top right), **c** Residual error, (bottom left), **d** The velocity of the forearm calculated by integrating the processed acceleration (bottom right)

Figure 6 (left and right) are the spectra for the left and right arms respectively. For clarity A2 was multiplied by 10^2 and A3 was multiplied by 10^4 so that the data sets do not obscure each other in the plot. In Fig. 6, the frequency peak at 9 Hz on the left arm is clearly visible in A3 although this was not evident in the movement without weights.

The speed of the arm movement appears as the lowest frequency peak. This is 0.3 Hz and 0.45 Hz, respectively, and was found to be twice at this fundamental frequency for A3. This is because the A3 sensor outputs positive values both above and below the horizontal plane. The ratio of the peak high oscillation frequency and the fundamental frequency are a measure of the number of unstable shoulder adjustments that occur during a single arm lift. Thus, for the left arm, there are approximately 37 steps but no equivalent effect in the right arm spectra. For all arm lifts with weights, the result was 35.2 ± 3.2 steps.

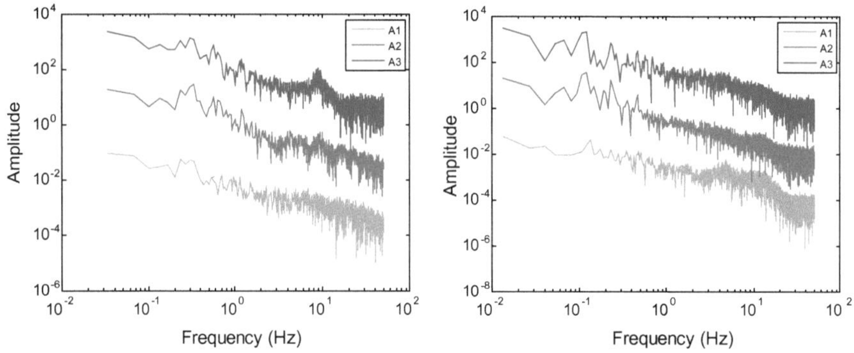

Fig. 6 Arm acceleration spectra for 3 kg weights on each arm for participant 1. Left arm (left); Right arm (right). The vibration at 9 Hz in the left arm is clearly evident but is not visible in the right arm spectrum

4 Conclusion and Further Work

Two different artefacts from acceleration sensor measurements were studied. A straight, horizontal and boxing jab with a wrist-mounted accelerometer into a wall-mounted bag included three different time sections: the acceleration, the rapid deceleration and oscillations of the sensor resulting from additional inertial movements of the sensor because of the skin elasticity. The frequency of the oscillations was found to be dependent on the weight of the sensor. A simple mathematical technique was used to remove the oscillation making the determination of the hand velocity straightforward. The technique can only be applied, if there is a clear understanding of the characteristic oscillation period for that skin surface and that sensor. There is an opportunity to measure this frequency before measurements are made in a realistic boxing situation. The analysis also provides a technique to quantify the method of attaching the sensor to the human body. This might prove a useful tool as placement of sensor on body locations with minimal soft tissue is the current strategy to minimize vibration artefacts.

A forward lift of a hand-held weight in front of the body without bending the elbow was measured using a wrist-mounted accelerometer. A comparison of the left and right arms of the participant who had a surgical rotator cuff repair to the left shoulder 4 years ago showed vibrations at 9 Hz. The accelerometer placement during this type of exercise could augment or replace complex and expensive three-dimensional computer-aided tomography, arthroscopic measurements, other imaging techniques and invasive explorations where the observation of dynamic movement is not possible and ligament imaging is usually not viable [10, 11]. This is a quantitative measure that could augment the questionnaire used to track non-traumatic shoulder instability over long periods of time [12], the longitudinal measurement of the shoulder instability index [13] and by comparing the acceleration recordings with internally generated audio sounds [14].

Acknowledgements The authors wish to thank the participants for their assistance and involvement in the project. Thanks also to Ray Leadbetter from Sabel labs for assistance with the instrumentation.

References

1. Cleland, I., Nugent, C. D., Finlay, D. D., Burns, W. P., Bougourd, J., Stevens, K., et al. (2012). Effects of accelerometer coupling on step counting accuracy in healthy older adults. *Health Technology, 2,* 259–270.
2. Thiel, D. V., Quandt, J., Carter, S. J. L., & Moyle, G. (2014). Accelerometer based performance assessment of basic routines in classical ballet. *Procedia Engineering, 72,* 14–19.
3. Stabb, W., Hottowitz, R., & Sohns, C. M. D. (2014). Sohns, J.M.MD., Gilbert, F.MD., Menke, J.MD., Niklas, A.MD., and Lotz, J.MD., "Accelerometer and gyroscope based gait analysis using spectral analysis of patients with osteoarthritis of the knee", Journal of Physical Thera py. *Science, 26,* 997–1002.
4. Liu, K., Liu, T., Shibata, K., Inoue, Y., & Zheng, R. (2009). Novel approach to ambulatory assessment of human segmental orientation on a wearable sensor system. *Journal of Biomechanics, 42,* 2747–2752.
5. Wakeling, J. M., & Nigg, B. M. (2001). Soft-tissue vibrations in the quadriceps measured with skin mounted transducers. *Journal of Biomechanics, 34,* 539–543.
6. Fratini, A., Cesarelli, M., Bifulco, P., & Romano, R. (2009). Relevance of motion artefact in electromyography recordings during vibration treatment. *Journal of Electromyography and Kinesiology, 19,* 710–718.
7. Kimm, D, & Thiel, D. V. (2015). Hand speed measurements in boxing. *Procedia Engineering, 112,* 502–506.
8. Espinosa, H. G., Lee, J., & James, D. A. (2015). The inertial sensor: A base platform for wider adoption in sports science applications. *Journal of Fitness Research, 4,* 13–20.
9. Thiel, D. V., Espinosa, H., Davis, G. M., Dylke, E., Foroughi, N., amd Kilbreath, L. (2013). Arm movement: The effect of obesity on a ctive lifestyles. *Procedia Engineering, 60,* 182–187.
10. Chuang, T.-Y., Adams, C. R., & Burkhart, S. S. (2008). Use of preoperative three-dimensional computers tomography to quantify glenoid bone loss in shoulder instability. *Arthroscopy: The Journal of Arthroscopic and Related Surgery, 24,* 376–382.
11. Burkhart, S. S., DeBeer, J. F., Tehrany, A. M., & Parten, P. M. (2002). Quantifying glenoid bone loss arthroscopically in shoulder instability. *Arthroscopy: The Journal of Arthroscope and Related Surgery, 18,* 488–491.
12. Dawson, J., Fitzpatrick, R., & Carr, A. (1999). The assement of shoulder instability: The development and validation of a questionaire. *The Bone and Joint Journal, 81-B,* 420–426.
13. Kirkley, A., Griffin, S., McLintock, H., & Ng, L. (1998). The development and evaluation of a disease-specific quality of life measurement tool for shoulder instability. *The American Journal of Sports Medicine, 26,* 764–772.
14. Espinosa, H. G., McDonough, E., & Thiel, D. V. (2016). Non-traumatic shoulder instability measurements from accelerometer and audio records. *Journal of Fitness Research, 5,* 11–13.

Biomechanics Analysis of Sepak Takraw Tekong Serves via Depth Camera Motion Capture System

Muhammad Zulhilmi Kaharuddin, Siti Badriah Khairu Razak, Mohamed Shawal Abd Rahman, Wee Chang An, Muhammad Ikram Kushairi and Mohd Zamani Ngali

Abstract This study is focussing on the biomechanics analysis of the *Sepak Takraw Tekong* pre-serve motion. To date, most biomechanics analysis in sports is using standard three-dimensional video camera recording as the motion capture system. In this project, an in-house marker-less three-dimensional (3D) Depth Camera motion capture system is introduced. Kinect for Xbox 360 is chosen as the 3D Depth Camera motion capture sensor. Two sensors are used for duplex motion capture system arrangement since a recent study has proven that duplex system provides better results as compared to single Kinect motion capture system. These sensors have the capability to capture motions at 30 fps and categorized as marker-less skeleton tracking system. A server or known as *Tekong* in *Sepak Takraw* sport was chosen as test subject and his motion during performing the pre-serve motion was analysed. The serves motion of *Tekong* was captured during training session. Total serves attempt captured were 15 attempts. Data collected from the Kinect was then analysed using Matrix Laboratory software. The parameters analysed were foot-to-foot distance and both knee relative angle at pre-serve phase. The mean value and standard deviation for foot-to-foot distance in this study are 0.58 and 0.19 m. For the biomechanics analysis on the sample chosen, the mean value for relative angle for right knee was 148.72° whereas left knee relative angle was 160.89°. Speed of the ball being served also recorded in this study. From the statistical analysis, the result shows that the foot-to-foot distance and relative angle of the knee during pre-serve phase have an effect on the speed of the ball being served. From the outcomes of the study, it is suggested to use Kinect as the motion capture system since it can provide marker-less which is unable to comfort the subject to be examined, and also it can provide data in three dimensional. But, improvement in the specification of Kinect sensor used was highly recommended.

M. Z. Kaharuddin (✉) · S. B. K. Razak · M. S. A. Rahman · W. C. An · M. I. Kushairi
M. Z. Ngali
Department of Mechanical, Faculty of Mechanical and Manufacturing Engineering, Universiti Tun Hussein Onn Malaysia, Parit Raja, Malaysia
e-mail: zulhilmikaharuddin.psrc1314@gmail.com

N. Sulaiman et al. (eds.), *Proceedings of the 3rd International Colloquium on Sports Science, Exercise, Engineering and Technology*,
https://doi.org/10.1007/978-981-10-6772-3_15

Keywords Biomechanics · *Tekong* · *Sepak takraw* · 3D depth camera · Kinect Xbox 360° · Pre-serve · Foot-to-foot · Relative knee angle · Speed · Speed radar gun

1 Introduction

The word biomechanics are from the prefix bio and the root word mechanics. The prefix bio visualizes that biomechanics has something to do with the analysis of force and their effects on the living systems. The root word mechanics indicates that biomechanics has something to do with the analysis of force and their effects. Therefore, the term biomechanics is the study of the structure and function of biological system by means of the methods of mechanics [1]. Recently, biomechanics becomes one of the vital fields in athlete performance evaluation. Biomechanics is described as the implementation of mechanical laws to living things or structure, specifically to the human body locomotor system [2].

Sepak takraw is Thai for 'kick' and 'woven ball', which defines the gist of this team sport that is known today as the international ball game of Southeast Asia [3]. The name comes from the combination of the Malay word *Sepak* (literally means as 'kick' or 'smash') and the Thai word *Takraw* (the original rattan ball used in the sport) [4]. *Sepak takraw* is a sport which is played in court. *Sepak Takraw* is a sport which is played in court which is 13.4 m long and 6.1 m width. The players tossing a rattan ball over a net barrier, which is played by two opposing teams consists of three players for each team. They can use any part of the body except their hands.

In modern *Sepak takraw*, *Tekong* plays a vital role. The server's role in *Sepak takraw* sport is crucial to the team's overall performance because a good service is sometimes sufficient in starting and terminating rallies in order to gain a point. Serving is viewed as *Sepak takraw* first measurement of offence because of its essentialness [5]. The serve is the first attacking strike in the modern game of *Sepak takraw*. Notational analysis conducted by National Sports Institute of Malaysia has shown that a good serve contributes to scoring points and winning the game [6]. *Tekong's* function in this sport is to serve the *Sepak Takraw* ball towards the opponent across the net barrier by using their legs. It required high skill level to perform the best serve in *Sepak Takraw* sport.

Therefore, the purpose of this project is to conduct the biomechanics analysis on the *Sepak Takraw Tekong* serve by using depth camera motion capture system. Motion capture system that has been used to capture the *Sepak Takraw Tekong* motion is the Kinect Xbox 360°. The Kinect is a novel motion analysis tool which allows markerless motion tracking at very low cost. The Kinect could be used for qualitative analysis in situations where ecological validity is of greater importance than the accuracy of measurement. This could include certain applications in coaching, teaching or clinical practice [7]. The serves motion of University Tun Hussein Onn (UTHM) *Sepak Takraw* players were captured during training sessions. The movement of the selected joint of the *tekong's* body during serving the *Sepak Takraw* ball was captured using Kinect Xbox 360° sensor. There are two different techniques that have been implemented to capture human motion, that is, marker and marker-less technique. The technique have been used in this study is marker-less. The Kinect should enable to capture human body joint motion and then retrieve the data for joint movement. The data are retrieved in three-dimensional (3D) axis which is x, y and z. From the previous study, there is a possibility to adopt marker-less motion capture technique in the biomechanics analysis. Moreover, it is suitable to implement since it can provide the reduction in preparation time of the subjects and the absence of markers that could modify the naturalness of a subject's movement [8]. Selected kinematic variables were calculated using data collected from Kinect sensor that linked to the Kinect for Windows software. The 3D data for 20 body joint captured from the Kinect is then processed using Matrix Laboratory (MATLAB) software. Then, the statistical analysis method is also used to define the most relevant kinematic factor that affects *Tekong* performance. The experiment is performed during the *Sepak Takraw* training session.

2 Method

2.1 Selected a Equipment

The study involved the analysis using motion capture using Kinect Xbox 360° sensor and software on the UTHM *Sepak Takraw Tekong*. Motion of one sample (*Tekong*/server) was captured in the training sessions. The performance of *Tekong* is measured from the consistency, ball placement and surely the speed of the ball being serves. Therefore, the parameter that needs to be recorded in the study is how fast the ball being served together with the consistency of the serve. The consistency of the serve in this study is referred to many trials of serve that successfully passes the net barrier and not going out from the standard court line dimension. There are several methods that can be used to measure the speed of the ball that is by using radar gun [9] or mathematical calculation from the data obtained by motion capture system. Fifteen (15) number of serves attempt were performed and recorded. The data of *Tekong* body joint motion were retrieved by using Kinect Xbox 360° sensor with resolution 640 × 480 and 30 frame rate per second (fps). Data collected from Kinect Xbox 360° were processed and analysed using MATLAB software. The analysis on joint motion was performed in three major axis which is x, y and z (3D). The collected 3D data for respective joint was used to perform kinematic analysis on the selected sample body joint. The selected parameters that have been analysed were the foot-to-foot distance and relative angle of the knee during pre-serve stage. The kinematic data for joint motion and motion pattern of selected joint were correlated with the speed of the ball using statistical analysis method.

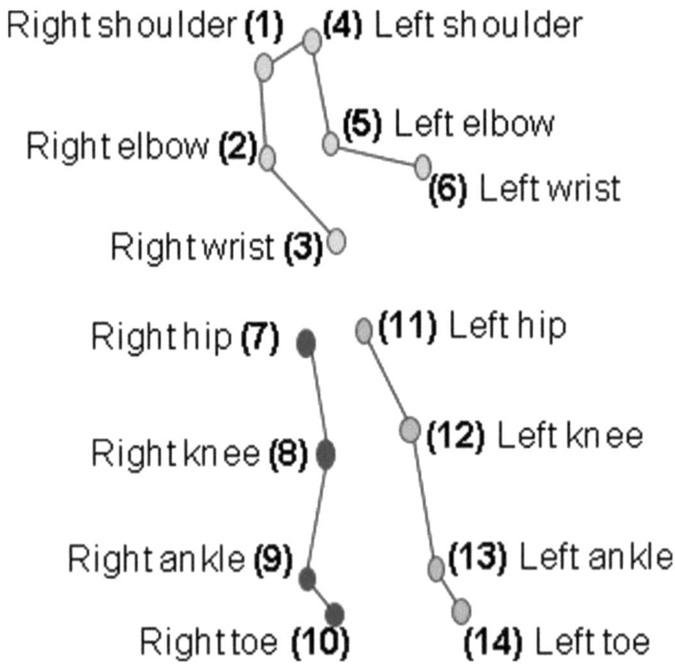

Fig. 1 Simplified skeleton model by MATLAB software

3 Results and Discussion

The skeleton model has been simplified using MATLAB software by processing data from Kinect Xbox 360°. The reason for simplification is to make the analysis procedure for selected joint easier. The simplified skeleton model for 14 selected joints by MATLAB software is shown in Fig. 1.

From the simplified skeleton model as shown in Fig. 1, the motion pattern of the *Tekong* during pre-serve phase can be observed. The sample of motion pattern of the *Tekong* during pre-serve phase is as shown in Fig. 2.

Parameters that have been analysed were foot-to-foot Euclidean distance and relative angle of knee during pre-serve phase. Foot-to-foot distance, d was calculated using the following formula:

$$d = \sqrt{(x_1 - x_2)^2 + (y_1 - y_2)^2 + (z_1 - z_2)^2} \quad (1)$$

where x_1, y_1 and z_1 are the 3D coordinates of right ankle of *Tekong* during pre-serve phase and x_2, y_2 and z_2 are the left ankle 3D coordinates of *Tekong* during pre-serve phase.

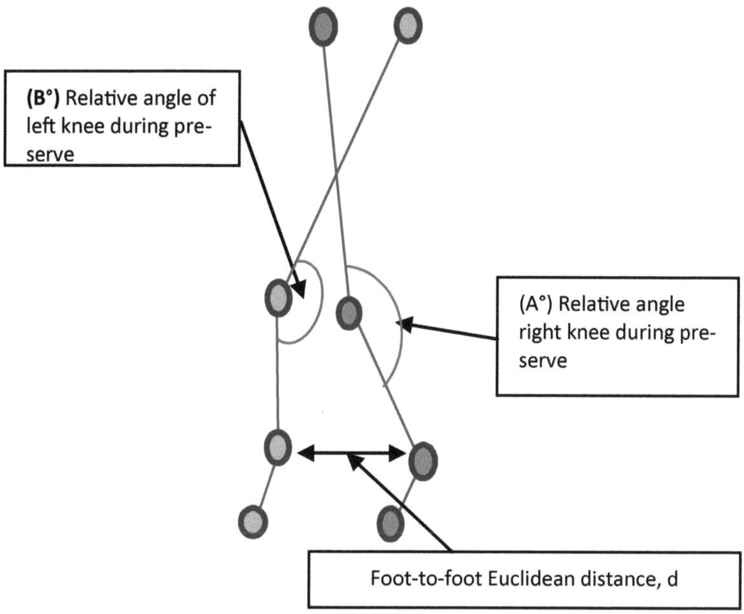

Fig. 2 Sample of *Tekong* motion pattern of right and left foot during pre-serve

Then, relative angle for knee, *(A, B)*, where *A* for right knee and *B* for left knee relative angle were calculated using cosine rule formula. These rules have been used due to triangular-shaped motion produced by connecting three jointd which are hip, knee and ankle joint of the *Tekong*. The triangular shape is produced from the pattern motion of *Tekong* as shown in Fig. 3.

$$\cos(A, B) = \frac{b^2 + c^2 - a^2}{2bc} \tag{2}$$

These trigonometric function and equation have been used in MATLAB programming. Data for 15 serve attempts of the selected parameters were calculated using MATLAB software. The speed of the ball served by *Tekong* and the serve outcomes whether in or out is also recorded. All of the data collected in the experiment were tabulated as shown in Table 1

From the data tabulated, the statistical analysis has been performed. The graph were plotted to visualize the correlation between parameter selected with the speed of the ball being served and the motion pattern of the sample during pre-serve phase. The graph plotted is shown in Figs. 4, 5, 6 and 7.

From the graph that have been plotted, Fig. 4 indicated that there is a correlation between foot-to-foot Euclidean distance, d with the speed of the ball being served. The graph shows that when the foot-to-foot distance is larger, the value for speed recorded is also faster. Then, from the graph in Figs. 5 and 6, it indicated that both

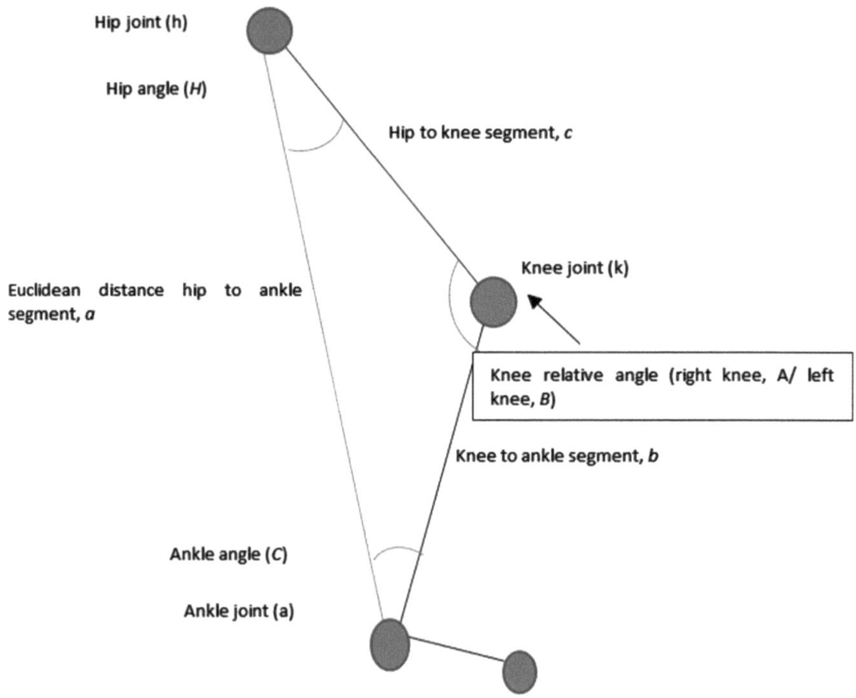

Fig. 3 Triangular shape produced from the pattern motion of *Tekong*

knee relative angles have contributed significant effect on the speed of the ball being served. From the comparison between the slope of both graph, it shows that right knee relative angle ($A°$) provided larger contribution on speed of the ball compared to left knee relative angle ($B°$). From Fig. 7, the graph that was plotted indicated that the movement of left knee of the sample is less consistent from the mean value due to large value of standard variation produced.

The analysis on the ball speed, s (km/h) recorded also have been performed. In this experiment, the fastest speed recorded was 69 km/h. The data tabulate shows that there was large deviation of serve speed for 15 serve attempts compared to mean speed. This situation indicated that the server also has less consistency in maintaining speed of the ball being served. Then, the serve outcome was also recorded. Out of the 15 serves, the sample manages to serve 10 attempts perfectly. These values are represented by 66.67% consistency level in the state of serve outcome.

Table 1 Summary kinematics data collected

Serve attempt	Foot-to-foot distance, d (m)	Relative angle for right knee, *A*(°)	Relative angle for left knee, *B*(°)	Serve outcome	Ball speed, s (km/h) (±1 km/h)
1	1.07	147.2	153.52	out	69
2	0.48	133.3	172.09	out	56
3	0.71	174.5	170.67	in	62
4	0.55	160.6	170.31	out	58
5	0.48	132.1	105.46	out	58
6	0.37	140.6	169.28	in	31
7	0.40	125.9	174.08	in	52
8	0.46	139.1	168.94	out	50
9	0.52	140.1	161.64	in	58
10	0.73	165.8	153.78	in	65
11	0.50	149.3	165.00	in	59
12	0.89	150.7	163.85	in	65
13	0.46	157.1	160.61	in	53
14	0.58	172.2	163.22	in	61
15	0.53	142.4	169.25	in	61
Mean	0.58	148.70	160.89	–	57.20
SD	0.19	14.70	17.20	–	8.87

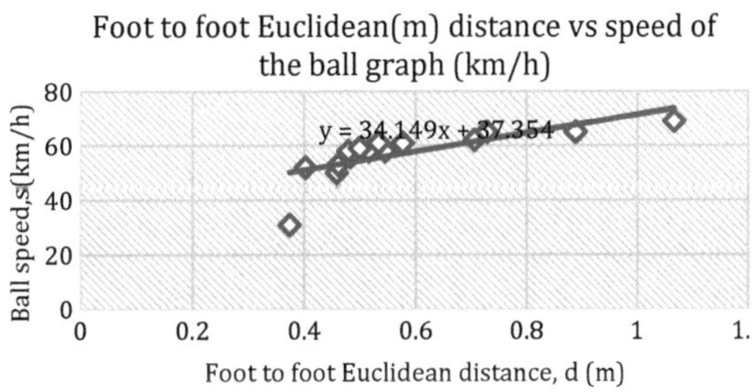

Fig. 4 Graph of foot-to-foot Euclidean, d (m) distance versus speed of the ball graph, s (km/h)

4 Conclusion

From the analysis and discussion, the results are obtained. The conclusion for this project is, there are three major kinematic parameters that have been analysed in this project which is the foot-to-foot distance at pre-serve or extension phase, relative angle of the knee at pre-serve or extension phase and speed of the ball being served by the *Tekong*. From the analysis that has been conducted, the general conclusion that can be made is the pre-serve motion technique can significantly affect the speed

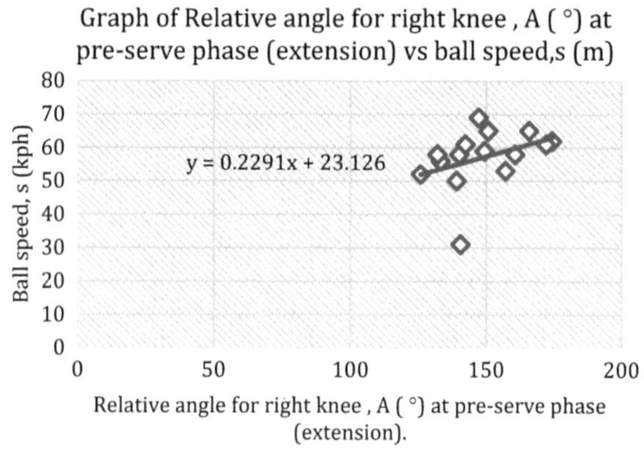

Fig. 5 Graph of relative angle for right knee, $A(°)$ at pre-serve phase (extension) versus ball speed, s (m)

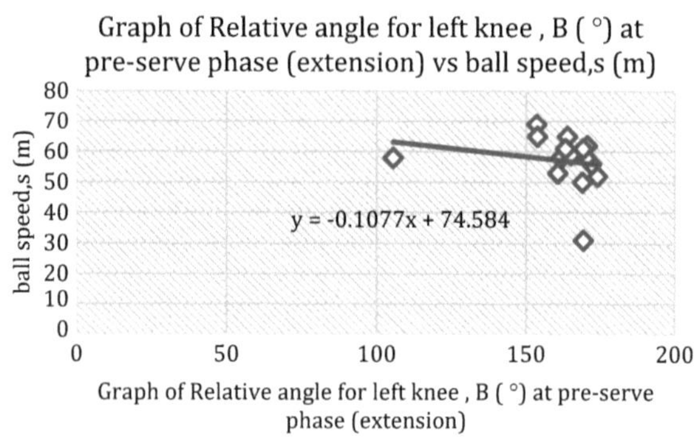

Fig. 6 Graph of relative angle for left knee, $B(°)$ at pre-serve phase (extension) versus ball speed, s (m)

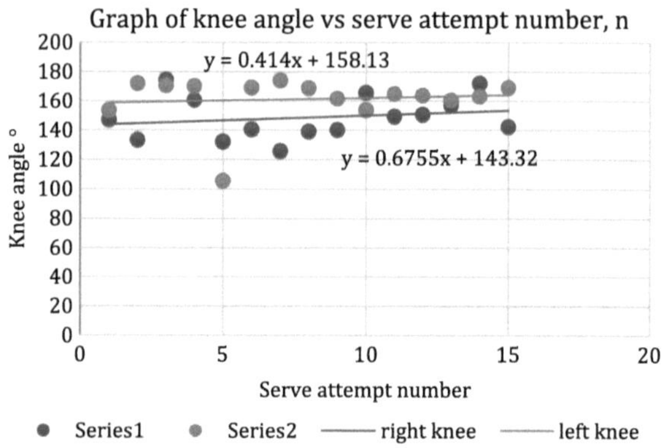

Fig. 7 Graph of knee angle ° versus serve attempt, n

of the ball served. This statement can be proven via referring to the data that were tabulated. From the data of the lowest serve speed recorded that is at 31 km/h, the foot-to-foot distance during pre-serve motion is the lowest among the other data.

Whereas, the data also shows that when the *Tekong* extends, the foot at the highest distance which is at 1.07 m, the *Tekong* able to serve at highest speed is recorded that is 69 km/h. For this biomechanics analysis, the mean value for foot-to-foot distance for the sample selected is 0.58 m and the standard deviation is 0.19 m.

Then, for the relative knee angle analysis, the data tabulated and graph plotted indicated that there are also significant effects of knee bend level towards the ball speed being served. From the comparison of both knee angles, the data trend shows that the right knee of the sample bends more compared to left knee during pre-serve phase. The angle of bend is also known as flexion angle. It is important for *Tekong* to minimize the flexion angle in order to straighten the served leg and serve the ball at higher point. For the biomechanics analysis on the sample chosen, the mean value for relative angle for right knee is 148.70°, whereas left knee relative angle is 160.89°. The standard deviation data shows that left knee relatives angle contribute larger variation data compared to right knee.

The additional parameter that has been analysed throughout the experiment is the speed of the ball being served. From the data tabulated, the highest ball speed recorded is 69 km/h. The mean value for ball speed is 57.2 km/h. The standard deviation for ball speed is 8.87 km/h. It can be claimed that the sample chosen served the ball with low consistency level.

Therefore, the conclusion that can be made from the biomechanics analysis of the *Sepak Takraw Tekong* serve motion is the kinematic parameter selected which are foot-to-foot Euclidean distance and relative angle of the knee have contributed towards the overall *Tekong* performance. It is also proven that the data retrieved from the Kinect Xbox 360° can be used in order to perform the analysis on the pre-serve technique of the *Tekong* sample.

Acknowledgements Author would like to thank the University Tun Hussein Onn Malaysia for giving the opportunity and provide sufficient facilities to run experiment regarding Biomechanics analysis on *Sepak takraw Tekong* pre-serve motion and this research is supported by the Fundamental Research Grant Scheme (FRGS), Grant No: FRGS Vot 1543 by Ministry of Education, Malaysia.

References

1. Hatze, H. (2015). The meaning of the term "biomechanics". *Journal of Biomechanics, 7*(2), 189–190. https://doi.org/10.1016/0021-9290(74)90060-8.
2. Garner, B. (2003). Joint anatomy and basic biomechanics. *Joint Anatomy, 568*.
3. Crego, R. (2003). *Sports and games throughout history: Sports and games of the 18th and 19th centuries*. Westport, London: Greenwood Press.
4. Federation, I. S. T. (n.d.). What is sepak takraw. Retrieved November 22, 2015, from http://www.sepaktakraw.org/about-istaf/how-to-play-the-game/.
5. Mohamad Razali, A., Saidon, A., & Pathmanathan, K. S. (2012). Visual perception of kuda and sila service. *Movement, Health and Exercise, 1*(1), 26–37.
6. Jamal, F. (2009). An analysis on Tekong service between two Malaysian Tekong in world championship 2009.
7. Simon Choppin, J. W. (2012). marker-less tracking of human movement using Microsoft kinect. In *30 International Conference on Biomechanics in Sports* (Vol. 86, pp. 231–234).

8. Ceseracciu, E., Sawacha, Z., & Cobelli, C. (2014). Comparison of markerless and marker-based motion capture technologies through simultaneous data collection during gait: Proof of concept. *PLoS ONE, 9*(3), e87640. https://doi.org/10.1371/journal.pone.0087640.
9. Hamdan, N., Suwarganda, E., & Wilson, B. (2012). Factors correlated with sepak takraw serve speed. In *30th Annual Conference of Biomechanics in Sports* (Vol. 186, pp. 413–416).

The Relationship Between Self-efficacy (Online Version) and 3 m Golf Putting Distance: A Pilot Study

Mazlan Ismail

Abstract The aim of this pilot study is to determine the relationship between self-efficacy and putting performance of the golfers. Hundred and twenty-two male golfers participated in this study in three different selected golf clubs. The online version of golf putting self-efficacy was developed in this study based on the written version created by the previous study. The 3 m distance was used as a measure in this study. To achieve the objective of this study, the golfers were required to rate their belief right before the putting test. The results showed no significant relationship between self-efficacy and putting performance from the 3 m distance. Contrary to the previous research where golfers perceived shorter distance is the hardest to perform psychologically. Future research should compare between short- and long-distance putting, if there is any association between distance and self-efficacy of the golfers.

Keywords E-putting self-efficacy · 3 m distance · Golf putting

1 Introduction

Golf is a game that you play against yourself and is popular among researchers [1–4]. Previous research found that pressure influences golfer's strokes routine based on kinematics and muscular factors [2]. Hence, it is clear that putting is harder to execute than driving and chipping [5]. Additionally, golfers feel more pressure by the putting green than by the strokes technique [6].

The relationship between athlete's self-efficacy and sport performance has become more popular among sports psychologist [7]. The efficacy influences athlete's thoughts and responsive reactions (e.g., worries, goals, attributions) and behavior (e.g., tenacity, selections, determination) [8]. For example, this can be observed when a golfer decides on what club to use based on their skills [9].

M. Ismail (✉)
Faculty of Sports Science and Recreation, Universiti Teknologi MARA Negeri Sembilan, Kuala Pilah, Malaysia
e-mail: mazlan.healthygeneration@gmail.com

N. Sulaiman et al. (eds.), *Proceedings of the 3rd International Colloquium on Sports Science, Exercise, Engineering and Technology*,
https://doi.org/10.1007/978-981-10-6772-3_16

Previous studies suggested that there is strong, positive correlation between self-efficacy on male and female golfers and 6 feet distance putting performance [1, 7]. Previous researchers have developed written self-efficacy scale particularly to measure putting performance from the specific distance [1]. The results supported the perspectives of golfers where 6 feet distance was considered the hardest distance to perform psychologically. However, past researchers had mentioned that trainers need to find tools that are easy to excess during psychological training [10, 11]. Indirectly, the online version to measure the psychological performance needs to be developed and helps coaches during training. Additionally, past study used a 3 m length distance and this was identified as a distance golfers would expect to hole a putt [12]. Thus, the question remains if there is any significant relationship between self-efficacy (by using the online version measurement) of golfers and 3 m distance putting performance.

2 Method

2.1 *Participants*

Researcher assessed by putting the ability of 122 amateur male golfers aged range between 20 and 42 years (M = 29.88, SD = 8.48), and with playing experiences average (M = 4.74, SD = 2.67).

2.2 *Instrumentation*

Online Version of Putting Self-efficacy Scale (e-Putting Self-efficacy)

For the purpose of this study, the online version self-efficacy scale was developed based on the scale created by a previous study [1]. The idea to develop this online measurement was due to the fact that it is mobile and easy to use when practicing putting at the actual green [10, 11]. As stated by the previous researchers, the written version was followed by a set of guidelines to construct the golf putting self-efficacy scale [1, 10]. The measure based on this scale was task-specific, arranged hierarchically to signify the increasing level of complexity, and concordance with the measure in performance [12–14]. Also, the homogeneity of items is not necessary since this scale is a hierarchical scale except the internal reliability [8], and the Cronbach's alpha coefficient was 0.86 [1]. Correlation results showed that there is positive relationship between self-efficacy and putting performance [1]. If the self-efficacy is increased there will be a corresponding increment in golfers putting scores.

The online version of golf putting self-efficacy scale also requires to rate the participants level of self-efficacy when putting from a specific distance [1]. Thus, the self-efficacy scores were based on the ranging scores (e.g., 10 putts; from 900

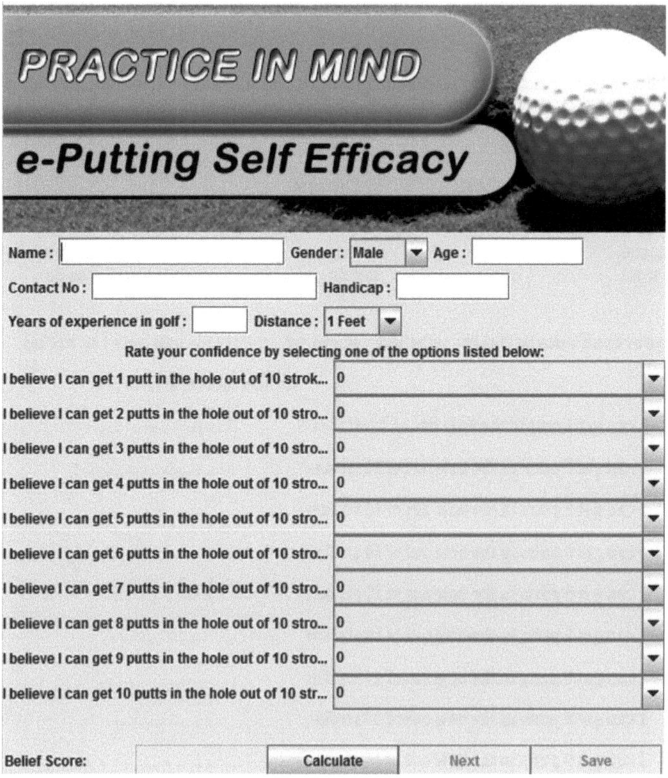

Fig. 1 The online version of golf putting self-efficacy assessment

to 1000 scores, 9 putts; from 800 to 899 scores, 8 putts; from 700 to 799 scores, 7 putts; from 600 to 699 scores, 6 putts; from 500 to 599 scores, 5 putts; from 400 to 499 scores, 4 putts; from 300 to 399 scores, 3 putts; from 200 to 299 scores, 2 putts; 100 to 199 scores, 1 putt; 0 to 99 scores) (see Fig. 1). Differences in written scores were based on a 100-point scale ranging from 0 (unable to do at all) 50 (moderately able), and 100 (highly certain able) (see Fig. 2).

Putting Scoring

A scoring system used by the previous study was five points given for putts in the hole, three points for any putts that stop at the lip of the hole but did not get in, two points awarded for putts that fell over the high side of the hole but did not get in, and one point for putts that fell by the low side or pull up short [1, 14]. Then, each participant was awarded a total score of a maximum of 50 points. Meanwhile, in the online version of putting self-efficacy, each participant was awarded a total score based only on the number putts in the hole or five points.

Putting Self-Efficacy

Rate your degree of confidence level by recording a number from 0 to 100 using the scale given below. Next, perform your skill and state your actual skill performance scores. Finally, evaluate the relationship between your confidence score and skill performance scores by recording as perfect, almost or try again.

PAR: ________ Distance: __________

0	10	20	30	40	50	60	70	80	90	100
Cannot do at all					moderately can do					highly certain can do

Rate your confidence level that you believe you can get...putt in the hole out of 10 strokes

	Confidence level (0-100)
I can get 1 putt in the hole out of 10 strokes	....
I can get 2 putts in the hole out of 10 strokes	....
I can get 3 putts in the hole out of 10 strokes	....
I can get 4 putts in the hole out of 10 strokes	....
I can get 5 putts in the hole out of 10 strokes	....
I can get 6 putts in the hole out of 10 strokes	....
I can get 7 putts in the hole out of 10 strokes	....
I can get 8 putts in the hole out of 10 strokes	....
I can get 9 putts in the hole out of 10 strokes	....
I can get 10 putts in the hole out of 10 strokes	....

Fig. 2 The written version of golf putting self-efficacy scale

2.3 Procedure

Three golf clubs were selected in this study. An informed consent was given to the golfers during the initial meeting. In addition, a professional golfer was appointed to avoid any biasness as well as to provide reference, especially in the technical aspect.

For the purpose of this study, each participant was asked to complete their online version of golf putting self-efficacy scale (e-putting self-efficacy). They have to key-in all the information such as name, gender, age, contact number, handicap level, and years of experience. They also need to key-in distance (3 m distance) at the distance column. In order to make the participants feel comfortable with their own putting technique, they were asked to use their own putter [1, 14]. Ten golf balls were also provided by the researchers. Consistent with the previous study, the present study uses 10 putting task [15]. The 10 putts were performed by all participants from a 3 m distance at the actual putting green (as approved by the professional golfer). Furthermore, for each golfer they need to key-in their putting score for each of the

Table 1 Pearson correlation between self-efficacy and 3 m distance putting performance in male amateur golfers ($N = 122$)

Variables	1	2
Self-efficacy	–	
3 m putting distance	0.01	–

strokes (10 strokes). The scoring system was based on the position of the scoring system recommended by the previous study. As such if a golfer had 5 points (hole out) for first stroke, he or she needs to key-in 5 points at the putting score performance column. Finally, the correlation score was calculated between the total scores of self-efficacy and total of putting scores (No. of 5 points score). Furthermore, all the scores were printed and analyzed by using SPSS in order to determine the relationship between the two variables.

3 Results and Discussion

The relationship between self-efficacy (as measured by e-putting self-efficacy) and 3 feet distance putting performance was investigated using Pearson product-moment correlation coefficient. Preliminary analyzes were performed to ensure no violation of the assumptions of normality, linearity and homoscedasticity. Table 1 showed that there was no correlation between two variables $r = 0.01$, $n = 122$, $p > 0.05$. This result suggested that there was no correlation between self-efficacy on male amateur golfers and 3 m distance putting performance. Contrary to the previous research where golfers perceived shorter distance is the hardest to perform psychologically [12].

4 Conclusion

Clearly, the results of this pilot study indicate that self-efficacy and 3 m distance putting performance was not significantly related between two variables, i.e., self-efficacy and 3 m putting distance. As supported by the previous researchers, golfers are more inclined to perform tasks that they feel efficacious in [1]. Indeed, a person who is highly efficacious in a particular task will initiate and continue in performing the task and vice versa [1, 16]. However, individuals who are weak in self-efficacy will try to avoid a challenging task [17].

Although the previous study has mentioned that the level of self-efficacy among male golfers is related to short distance particularly the 6 feet distance but not from the 3 m (10 feet) distance. This pilot study, therefore, allows a number of potential research questions. For instance, it was discovered that self-efficacy of the golfers when putting from 3 m distance is somewhat not related with personal feelings and believe when playing. This is inconsistent to the 6 feet distance study that revealed self-belief and confidence may occur during every competition, especially in golf when golfers putt from the short distance [6, 18]. However, future research should compare between short- and long-distance putting if there is any association between distance and self-efficacy of the golfers.

This study most importantly, suggested that the online version or e-Putting self-efficacy is useful to measure the psychological state of the golfers during imagery training at the actual environment. Hence, coaches may use the same tool to measure other distances and explore the level of self-efficacy in a competition condition.

Acknowledgements The author would like to thank Healthy Generation Malaysia, the golf clubs involved, the coaches and players for the support and cooperation in completing this study.

References

1. Ismail, M., & Ramlee, W. (2016). The relationship between self-efficacy and 6 feet golf putting distance. In S. I. Ismail et al. (Eds.), *Proceedings of the 2nd International Colloquium on Sports Science, Exercise, Engineering and Technology 2015 (Icosseet 2015)* (pp. 247–252). Singapore: Springer. https://doi.org/10.1007/9789812876911126.
2. Cooke, A., Kavussanu, M., McIntyre, D., & Ring, C. (2010). Psychological, muscular, and kinematic factors mediate performance under pressure. *Psychophysiology, 47,* 1109–1118. https://doi.org/10.1111/1469-8986.2010.01021.
3. Hayslip, B., Petrie, T. A., MacIntire, M. M., & Jones, G. M. (2010). The influences of skill level, anxiety, and psychological skills use on amateur golfers' performance. *Journal of Applied Sport Psychology, 22,* 123–133. https://doi.org/10.1080/10413200903554281.
4. Roberts, R., & Turnbull, O. H. (2010). Putts that get missed on the right: In vestigating lateralized attentional biases and the nature of putting errors in golf. *Journal of Sport Science, 28,* 369–374. https://doi.org/10.1080/02640410903536467.
5. Hung, G. K. (2003). Effect of putting grip on eye and head movements during the golf putting stroke. *The Scientific World Journal, 3,* 122–137. https://doi.org/10.1100/2003.14.
6. Clark, R. D. I. I. I. (2004). Streakiness among professional golfers: Fact or fiction? *International Journal Sport Psychology, 34,* 63–79.
7. Ismail, M. (2014). Golf putting: Shorter putts are easier, is this really true. *International Journal of Enhanced Research in Educational Development, 2,* 22–27.
8. Feltz, D. L., Short, S. E., & Sullivan, P. J. (2008). *Self-efficacy in sport; research and strategies for working with athletes, teams, and coaches*. United States: Human Kinetics.
9. Bandura, A. (Ed.). (1997). *Self-efficacy: The exercise of control*. New York: Freeman.
10. Ismail, M. (2014). Practice in mind: Help to improve golf putting from the hardest distance. *International Journal of Enhanced Research in Educational Development, 2,* 7–12.
11. Smith, D., Wright, C. J., & Cantwell, C. (2008). Beating the bunker: The effect of PETTLEP imagery on golf bunker shot performance. *Research Quarterly for Exercise and Sport, 79,* 385–391.
12. Smith, D., & Holmes, P. (2004). The effect of imagery modality on golf putting performance. *Journal of Sport and Exercise Psychology, 26,* 385–395.
13. Moritz, S. E., Feltz, D. L., Fahrbach, K. R., & Mack, D. E. (2000). The relation of self-efficacy measures to sport performance: A meta analytic review. *Research Quarterly for Exercise and Sport, 71,* 280–294.

14. Cumming, J., Nordin, S. M., Horton, R., & Reynolds, S. (2006). Examining the direction of imagery and self talk on dart—Throwing performance and self-efficacy. *The Sport Psychologist, 20,* 257–274.
15. Nordin, S. M., & Cumming, J. (2005). Professional dancers describe their imagery: Where, when, what, why, and how. *The Sport Psychologist, 19,* 395–416.
16. Short, S. E., Bruggeman, J. M., Engel, S. G., Marback, T. L., Wang, A. W., & Short, M. W. (2002). The effect of imagery function and imagery direction on self-efficacy and performance on a golf putting task. *The Sports Psychologists, 16,* 48–67.
17. Ramsey, R., Cumming, J., & Edwards, M. G. (2008). Exploring a modified conceptualization of imagery direction and golf putting performance. *International Journal of Sport and Exercise Psychology, 6,* 207–223.
18. Lane, A. M., & Jarrett, H. (2005). Mood changes following golf among senior recreational players. *Journal of Sport Science and Medicine, 4,* 47–51.

Association of Spectators Based Brand Equity (SBBE) and Fans Attendance: A Case of Selangor Football Fans

Ong Tah Fatt and Muhammad Safuan Bin Aziz

Abstract The concept of brand equity is often used to analyze how a brand can add value to a product or service. Brand researchers have developed several conceptualizations of brands equity in understanding consumer behavior in different industry. However, there is a paucity of research utilizing the brand equity concepts in comprehending sports consumer in supporting successful sports team. Thus, the present study aims to examine the relationship of brand association and brand awareness with the attendance of Selangor football fans, utilizing the Spectator-Based Brands Equity (SBBE) model. Convenience sampling technique was used to involve 150 Selangor football fans attending football matches in Stadium Shah Alam, as respondents. Data were collected through self-administered questionnaire. In brand association, three dimensions with the highest mean are team history ($M = 5.99$), followed by commitment ($M = 5.81$), and team success ($M = 5.68$). As for brands awareness, identification ($M = 5.78$) has a higher mean than internalization ($M = 5.37$). Both brand awareness ($r = 0.638, p < 0.001$), and brand association ($r = 0.572, p < 0.001$) are significantly related to attendance of fans. In conclusion, brands awareness plays a more important role in influencing the attendance of Selangor Football fans. Several suggestions are discussed and recommended for Selangor FA to attract more supporters and sustain fans attendance.

Keywords Brand equity · Brand association · Brand awareness

1 Introduction

Football is the most popular sport not only in Malaysia but also in the world. Professional football (soccer) is recognized as the sport with the most followers (and viewers) on the planet [1]. The most popular football league tournament among Malaysian spectators is the Malaysia Cup, which is managed by Malaysia Football

O. T. Fatt (✉) · M. S. B. Aziz
Faculty of Sports Science and Recreation, Universiti Teknologi MARA, Shah Alam, Malaysia
e-mail: ong278@salam.uitm.edu.my

N. Sulaiman et al. (eds.), *Proceedings of the 3rd International Colloquium on Sports Science, Exercise, Engineering and Technology*,
https://doi.org/10.1007/978-981-10-6772-3_17

Association (FAM). Malaysia Cup is contested by the top sixteen football clubs in the country. Selangor Football Association (SFA) has won this Cup for 23 times (until 2015) since the cup have been introduced in 1921, and is the most successful team in Malaysia. The Selangor team has a large number of supporters and fans that indirectly contributed to the success of the team over the years. In Malaysia Cup, football spectators are increasing every season. It reflects a good development process of Malaysian football industry. To enhance positive behavior and attract more people to support the teams, the associations and the management need to improve their quality in all aspects to ensure consumers or spectators will continue buying their products and support the football teams. This is one of the essential marketing strategies that need to be emphasized in order to warrant that the product meets consumer's choice.

Building a strong brand in consumer's mind is one of the most important ways for marketers to achieve the goal of product and brand management [2]. Many companies use the concept of brand equity to understand the success of their product among the consumers. Brand researchers have developed several conceptualizations of brands and how brands affect consumer behavior of current and future purchases [3]. Earlier models—such as Aaker's [4] brand equity model and Keller's [5] customer-based brand equity model—have focused heavily on how consumer perceive and evaluate brands by investigating certain knowledge structures such as loyalty, awareness, and perceive quality [4]. Ross [6] has introduced the Spectator-Based Brands Equity (SBBE) model with an attempt to comprehend how consumer evaluates and choose a sports service. The SBBE model consists of two components: brand association and brand awareness. In relation to a sports team, this could be an evaluation tool for the marketers or sport manager to evaluate and develop strategies which can be beneficial in developing an effective brands equity for the team. It is also possible to identify the consumer choice when they support their team based on the elements outlined in brand association and brands awareness.

Many previous researches on brand equity study have focused on the sports products such as replacement on brand lining [7]. There is a paucity of research on brands equity in relation to spectators. Previous research by Ross et al. [8] has involved consumers from a National Basketball Association. There are also other studies using customer-based brand equity (CBBE) which focus on European Clubs, focusing on their corporate social responsibility (CSR) and engagement impact [1]. Though Spectator-Based Brand Equity (SBBE) model has been introduced in understanding professional soccer spectators in Portuguese league [9], the model has not been applied to the local Malaysian soccer spectators or fans. In terms of methodology, previous research has been conducted using online survey technique [9]. The researcher also suggested that collecting additional data at the actual stadium may contribute to a more representative sample of the club's fan base. Thus, the present study aims to examine the relationship between SBBE and attendance of sports fans involving Selangor football fans attending matches at the stadium. Specifically, the SBBE model examines the relationship of brands awareness and brands association toward the behavioral intention of Selangor football fans to watch matches of Selangor team at the stadium.

2 Methodology

2.1 Research Design

The survey method was used to gather information from the Selangor football fans involving close-ended questionnaire. A convenience sampling technique was utilized, whereby self-administered questionnaire was distributed to 150 Selangor fans who came to Stadium Shah Alam to support their team. Respondents were met onsite and were kindly asked to participate in the survey voluntarily. The researcher was present throughout the session to answer any question from the respondents, relating to the questionnaires. No time limit was placed on the respondents to complete the inventory. The age of the respondents in the study was at least 18 years old and above.

2.2 Instrumentation

For better comprehension of the respondents, the research questionnaire was prepared in both English and Malay language and was designed in four sections. The first section is about demographic background of the respondents, with questions related to gender, age, race, marital status, and occupation. The second section consists of 29 items, adapted from the Spectator-Based Brand Equity Scale [9], which measures the nine dimensions of brand association. The nine dimensions were brands mark, rivalry, social interaction, team history, team success, non-player personnel, stadium community, commitment, and stadium community. In the third section, eight items were used to measure brand awareness, which comprises two dimensions (identification and internalization). For the final section, behavioral intention of fans was measured by four items. Seven-point Likert scale was used for the entire questionnaire. A pilot study was conducted with 40 respondents to test the validity and reliability of the survey instrument. Cronbach's alpha reliability coefficient for brand association was $\alpha = 0.94$, brand awareness was $\alpha = 0.90$, and behavioral intention was $\alpha = 0.87$. The instrument was considered to have acceptable reliability, with alpha value greater than 0.70.

2.3 Data Analysis

Data analysis was conducted utilizing the Statistical Package for Social Sciences (SPSS Ver.21). For demographic variables (gender, age, race, marital status, and occupation) of respondents, descriptive statistics such as percentage, frequencies, mean, and ranking was used to explain and describe the findings. Pearson correlation was used to examine the strength of the relationship between the independent variables (brand association and brand awareness) and dependent variable (behavioral intention).

Table 1 Demographic profile of respondents

Demographic characteristics		Frequency (N) ($N = 150$)	Percentage (%)
Gender	Male	121	80.7
	Female	29	19.3
Age	18–24 years	54	36.0
	25–31 years	62	41.3
	32 years and above	34	22.7
Marital status	Single	103	68.7
	Married	47	31.3
Ethnicity	Malay	120	80.0
	Chinese	15	10.0
	Indian	15	10.0
Occupation	Public sector	26	17.3
	Private sector	37	24.7
	Self-employed	31	20.7
	Unemployed	10	6.7
	Student	46	30.6

3 Results and Discussion

3.1 *Demographic Profile of Respondents*

The results of the demographic background of respondents were summarized in Table 1. In terms of gender, majority of the respondents were male (80.7%), while 19.3% of the respondents were female.

With regards to age group, most of the respondents were young adults with 77.3% aged from 18 to 31 years old. Corresponding to marital status, single respondents (68.7%) were more than respondents who are married (31.3%). In terms of ethnicity, the majority of respondents were Malays (80%), followed by Chinese (10%), and Indian (10%). The result reflects the true composition of ethnic groups in Malaysia, whereby Malay is the dominant ethnic group followed by Chinese and Indian. Among the respondents, student has the highest percentage (30.6%), followed by fans from the private sectors (24.7%), and self-employed (20.7%).

3.2 *Dimensions of Brand Association and Brand Awareness*

Table 2 below depicts the descriptive statistic for nine dimensions of brands association that includes team history, commitment, rivalry, team success, brands mark, stadium community, organizational attributes, social interact, and non-player per-

Table 2 Descriptive statistics of brands association and brand awareness

Dimensions	Mean	Std. deviation	Rank
Brand association			
Team history	5.99	0.87	1
Commitment	5.81	0.81	2
Team success	5.68	0.83	3
Brand mark	5.66	0.35	4
Stadium community	5.64	0.52	5
Organizational attributes	5.52	0.84	6
Social interact	5.52	0.98	7
Non-player personnel	5.52	0.93	8
Rivalry	4.94	0.77	9
Overall mean	5.58	0.76	
Brand awareness			
Identification	5.78	1.08	1
Internalization	5.37	1.21	2
Overall mean	5.57	1.45	

sonnel. Team history dimension has the highest mean of 5.99 compared to other dimensions in brands association. As for brand awareness, there are two dimensions identification and internalization. Comparatively, identification dimension has the higher mean value ($m = 5.78$), than internalization dimension ($m = 5.37$)

3.3 *Correlation Between Dimensions of Brand Association, Brand Awareness, and Behavioral Intention*

From the result in Table 3, the highest correlation between the dimensions of brands associated with behavioral intention is "organizational attributes" ($r = 0.543$). The second and third highest dimensions are "social interaction" ($r = 0.537$), and "team success" ($r = 0.507$). For the dimensions of brand awareness, "internalization" ($r = 0.583$) has a higher correlation with behavioral intention compared to "identification" dimension ($r = 0.583$).

Among the dimensions of brands association, "organizational attributes" has the highest correlation with behavioral intention. Organizational attributes encompass values and behavior that contribute to the unique social and psychological environment of an organization. It was found that Selangor football team had regularly organized meetings and great session with their fans. This kind of program depicts the symbolic appreciation of the Selangor to their football fans for their all-time support. Biscaia et al. [3] has highlighted the importance of non-product-related attributes in sports organizations. McAlexander et al. [10] agree that implementing annual fan

Table 3 Correlations between dimensions of brands association, brand awareness and behavioral intention

Brands association	Behavioral intention	
	Pearson correlation	Sig. (2-tailed)
Organizational attributes	0.543**	0.000
Social interaction	0.537**	0.000
Team success	0.507**	0.000
Non-player personnel	0.492**	0.000
Rivalry	0.458**	0.000
Stadium community	0.419**	0.000
Brands mark	0.325**	0.000
Team history	0.309**	0.000
Commitment	0.295**	0.000
Brands awareness		
Internalization	0.583**	0.000
Identification	0.537**	0.000

**Correlation is significant at the 0.01 level (2-tailed)

satisfaction survey may prove to be crucial in designing marketing programs that strengthen fans' connection with the team, and positively influence their perception about the organizational attributes.

The "social interaction" dimension of brands association has the second highest correlation with behavioral intention. Social interaction is recognized as an exchange between two or more individuals and is a building block of society. In the current study, respondents indicated that the team offers a place to meet other people and spend times with friends. It means that many Selangor fans came to the stadium with their friends to build a positive relationship with others and provide opportunity interacting with their friends. Spectating alone would be boring even the match is interesting. This is related to the previous study that sharing the sports experience with other fans is another way of contributing to leveraging the sports team's brand [11]. The present finding is consistent with prior research by Bauer et al. [12], who found that social interaction was a significant predictor of brand association. Thus, suggesting that the experiential benefits are an important aspect of spectator's consumption experience.

"Team success" was the third highest related dimension with behavioral intention. Team success reflects the glory that the team has achieved in past and present, in which fans can remember and cherish. Selangor football team has proven its supremacy in the Malaysia Cup when the team had won the title for the 33rd time in 2015. With this unprecedented achievement, the team, in return, has showed their appreciation for the support they had received from the fans over the years. As highlighted by Ross et al. [8], the consistency of a winning team and team success can influence the brand association dimension.

In terms of correlation between brand awareness and behavioral intention, internalization dimension ($r = 0.583$) possess a stronger relationship than the "identification" dimension ($r = 0.560$). The present finding supports the previous study which found that identification and internalization are important components of brands association [9, 12]. For internalization, Ross et al. [8] noted that an individual's psychological connection with a team serves as a gage to his/her awareness of the sport brand. Keller [5] explained that brand awareness plays an important role in consumer decision making. In addition, Ross et al. [13] found that identification has a positive relation with social interaction. Their study suggested that by improving the quality of concession areas (e.g., partnerships with food companies that fans appreciate), the teams will boost opportunities for fans to socialize and consequently increase their levels of identification.

The present research findings support the spectator-based brand equity model that has been proposed by Ross [6]. Brand association and brand awareness are highly correlated with the behavioral intention of Selangor football fans' attendance to the stadium. Recognizing the important role of "organizational attributes" in influencing attendance of football fans, managers should consider to promote more meaningful community program to attract and show appreciation to their fans. Activities such as hosting charity programs to raise fund helping the community in need, natural disaster victims or underprivileged community should be promoted. Fulfilling the community social responsibility (CSR) can indirectly influence the perception of the fans toward the team and enhance its brand image. This could generate the sense of belonging of the fans, which further develop a feeling of strong internalization towards the team.

For future study, it is recommended that more variables could be attempted using the SBBE model. For example, studies on impact of outcome variables such as media contracts, media consumption, and merchandise purchase behavior can be investigated. In the present research, female respondents were very limited. Previous research has mentioned that spectators perception about sporting events tends to vary according to gender. Thus, the future study based on gender difference is recommended. In addition, using SBBE model to examine on other sports and their attendance could also be attempted.

References

1. Blumrodt, J., Bryson, D., & Flanagan, J. (2012). European football teams' CSR engagement impacts on customer-based brand equity. *Journal of Consumer Marketing, 29*(7), 482–493.
2. Kapferer, J. N. (2008). *The new strategic brand management: Creating and sustaining brand equity* (2nd ed.). London: d'Organization.

3. Esch, F. R., Langner, T., Schmitt, B. H., & Geus, P. (2006). Are brands forever? How brand knowledge and relationships affect current and future purchases. *Journal of Product & Brand Management, 15*(2), 98–105.
4. Aaker, D. A. (1991). *Managing brand equity: Capitalizing on the value of a brand name*. New York, NY: The Free Press.
5. Keller, K. L. (1993). Conceptualizing, measuring, and managing customer-based brand equity. *The Journal of Marketing*, 1–22.
6. Ross, S. D. (2006). A conceptual framework for understanding spectator-based brand equity. *Journal of Sport Management, 20*(1), 22.
7. Li, H., Jin, H., & Yuan, G. (2011). Research on brand equity of sports—Take the replacement of the brand lining as example. *Journal of Sustainable Development, 4*(1), 207.
8. Ross, S. D., Russell, K. C., & Bang, H. (2008). An empirical assessment of spectator-based brand equity. *Journal of Sport Management, 22*(3), 322.
9. Biscaia, R., Rosado, A., & Maroco, J. (2013). Spectator-based brand equity in professional soccer. *Sport Marketing Quarterly, 22*(1), 20–33.
10. McAlexander, J. H., Schouten, J. W., & Koenig, H. F. (2002). Building brand community. *Journal of marketing, 66*(1), 38–54.
11. Underwood, R., Bond, E., & Baer, R. (2001). Building service brands via social identity: Lessons from the sports marketplace. *Journal of Marketing Theory and Practice*, 1–13.
12. Bauer, H. H., Stokburger-Sauer, N. E., & Exler, S. (2008). Brand image and fan loyalty in professional sport team: A refined model and empirical assessment. *Journal of Sport Management, 22*(2), 205–226.
13. Ross, S. D., Walsh, P., & Maxwell, H. D. (2008). The impact of team identification on ice hockey brand associations. *International Journal of Sport Management and Marketing, 5*(1–2), 196–210.

Spectators Motivation and Satisfaction in Local Motorsport Event

Rezian-na Muhammed Kassim

Abstract Motivation and satisfaction have a big impact towards the motorsport event industry, but it depends on the spectator Furthermore, the local motorsport event has become an important industry in Malaysia but untapped market for scientific research. Therefore, this study examines the relationship between motivation and satisfaction of spectator of local Motorsport events. This study used the scale of Facet of motivation (SPEED) as a motivation factor. SPEED facets of motivation were comprised of socialization, performance, excitement, esteem, and diversion. For this study, an event intercept method was used with a total of 290 respondents. A valid questionnaire was analysed using Pearson Correlation. The results of the Pearson Correlation analysis showed that there is a positive correlation between SPEED facet of motivation and satisfaction. In addition, it was found that all hypotheses are accepted. Thus, this finding will be used as important information for future research and organizer.

Keywords Motivation · Satisfaction · Relation · Local motorsport event

1 Introduction

Nowadays, the spectator has become an important element in every sport event. They have been recognized as a most influent factor of every sport event as a spectators, are the key constituent of a sport organization's success [1]. Events are essentially subjective experiences that are difficult to measure, in which sportsmen and spectators alike are actually part of the action [2]. Sport events are important to spectators for many reasons [3]. This study objective was to applied SPEED Facet of Motivation and satisfaction. SPEED represents the variables of social, performance, excitement, esteem and diversion. SPEED is a compilation of core motivational studies and has been an established model to test the motivation behavioural of people which leads to

R. M. Kassim (✉)
Faculty of Sports Science and Recreation, Universiti Teknologi MARA, Shah Alam, Malaysia
e-mail: rezian@salam.uitm.edu.my

N. Sulaiman et al. (eds.), *Proceedings of the 3rd International Colloquium on Sports Science, Exercise, Engineering and Technology*,
https://doi.org/10.1007/978-981-10-6772-3_18

satisfaction [4], Therefore, satisfaction is an important indicator of for the organizer to recognize the needs of the spectator been accomplished [3]. Thus, marketer or organizer of local motorsport event will use the information in order to attract more spectators. Most of the studies have been focused on the main event rather than local event motorsport.

Thus, this study focuses on the local Malaysia event to gain better scientific information about motorsport of motivation and satisfaction. Moreover, motivation is an inner drive to behave or act in a certain manner [5]. So, there is the need to find out what motivates people to become spectators and what motivates them to travel to attend these events [6]. Additionally, limited research has addressed the satisfaction of sport spectator, with even fewer studies examining the determinants of this situation [7]. Spectators may be satisfied only when they truly believe that they have received value for their time and money [8].

2 Research Design

This study utilized the quantitative study whereby the independent variable consists of Speed Facet of motivation factor (socialization, performance, excitement, esteem and diversion) and dependent variables is satisfaction. The questionnaire was distributed by 20 numerators and the event intercept method was used to interview the spectator (n300) and 290 return back. From the last year data, the population of 20,000 spectators attended the local motorsport event, therefore 300 sample size justified enough for the population of 20,000 [9].

3 Results and Findings

3.1 Demographic Result

A total of 300 questionnaires were distributed, of which 290 were returned completed and met the requirement of the study, representing a net response rate of 96% out of which. Among the 290 respondents, 50.6% were males and 49.4%. As expected, a majority of them were Malaysian. Most of the respondents were aged from 31 to 40 years old with a rate of 47%. While the rest ranged between 20 and 30 years old (43%) and 50–64 years old stated 10%. The majority of the races were Malay with rates of 78%, Chinese and Indian stated 10 and 10% and other 2%

Factors and item	Mean	Std. deviation
1. *Socialization*		
To socialize with other	3.95	0.70
To talk with other	3.90	0.70
To interact with others	3.95	0.71
2. *Performance*		
Attraction of the performance	3.91	0.74
The beauty of the event	3.94	0.76
The nature of the event	3.88	0.73
3. *Excitement*		
The excitement surrounding the event	3.81	0.80
The event is very exciting	3.81	0.77
The excitement associated with the event	3.82	0.77
4. *Esteem*		
I am one when the team wins	4.07	0.81
I feel motivated when the team wins	4.11	0.75
I get a sense of accomplishment when the team wins	4.09	0.76
5. *Diversion*		
I can get away from tension	3.90	0.86
Provide break from daily routine	3.92	0.87
Provide distraction from everyday activities	3.89	0.89

3.2 Result of SPEED Factor in Satisfaction

For socialization, the item that has higher mean and standard deviation was 'To interact with other' ($M \pm$ SD) (3.95 ± 0.71). The item that has lowest mean and standard deviation was 'To talking with other' ($M \pm$ SD) (3.90 ± 0.70). For performance, the item that has higher mean and standard deviation of 'The beauty of the event' ($M \pm$ SD) (3.94 ± 0.76). The item that has lowest mean and standard deviation of 'The nature of the event' ($M \pm$ SD) (3.88 ± 0.73). For excitement, the item that has higher mean and standard deviation was 'The excitement associated with the event' ($M \pm$ SD) (3.82 ± 0.77). The item that has lowest mean and standard deviation was 'The event is very exciting' ($M \pm$ SD) (3.81 ± 0.77). For esteem, the item that has higher mean and standard deviation was 'I feel motivated when the team win' ($M \pm$ SD) (4.11 ± 0.75). The item that has lowest mean and standard deviation was 'I am one when the team win' ($M \pm$ SD) (4.07 ± 0.81). For diversion, the item that has higher mean and standard deviation was 'Provide break from daily routine' ($M \pm$ SD) (3.92 ± 0.87). The item that has lowest mean and standard deviation was 'Provide distraction from everyday activities' (M ± SD) (3.89 ± 0.89).

3.3 Satisfaction in SPEED Factor

Satisfaction items	Means	Std. deviation
Socialization • Meet new people	4.50	0.50
• Contact with other friends	3.93	0.70
Performance • Chance to learn new things	3.91	0.74
• Develop a new skill	4.20	0.54
Excitement • There are adventures and excitement	3.82	0.77
• Different uncertainty will happen	4.33	0.67
Esteem • Keep peace in mind	4.09	0.78
• Its benefits to the society	4.49	0.52
Diversion • Can escape from pressure	3.90	0.87
• Have a better understanding of myself	4.14	0.68

For the satisfaction of socialization, the item that has higher mean and standard deviation was 'Meet new people' ($M \pm$ SD) (4.50 ± 0.50). The item that has lowest mean and standard deviation was 'Contact with other' ($M \pm$ SD) (3.93 ± 0.70). For the satisfaction of performance, the item that has higher mean and standard deviation was 'Develop a new skill' ($M \pm$ SD) (4.20 ± 0.54). The item that has lowest mean and standard deviation was 'Chance to learn new things' ($M \pm$ SD) (3.91 ± 0.74). For the satisfaction of excitement, the item that has higher mean and standard deviation was 'Different uncertainty will happen' ($M \pm$ SD) (4.33 ± 0.67). The item that has lowest mean and standard deviation was 'The is adventure and excitement' ($M \pm$ SD) (3.82 ± 0.77). For the satisfaction of esteem, the item that has higher mean and standard deviation was 'Its benefits to the society' ($M \pm$ SD) (4.49 ± 0.52). The item that has lowest mean and standard deviation was 'Keep peace in mind' ($M \pm$ SD) (4.09 ± 0.780). For the satisfaction of diversion, the item that has higher mean and standard deviation was 'Have a better understanding of myself' ($M \pm$ SD) (4.14 ± 0.680).

3.4 *Result for Pearson Correlation of SPEED Factor and Satisfaction*

Factors	Satisfaction (*r*)	*P* value
Socialization	0.456[a]	0.00
Performance	0.328[a]	0.00
Excitement	0.460[a]	0.00
Esteem	0.294[a]	0.00
Diversion	0.370[a]	0.00

[a]Correlation is significant at the 0.01 level (2-tailed)

H1 = There is a significant relationship between socialization and satisfaction at the local Motorsport event. Socialization and satisfaction of the spectator were moderately correlated, ($r = 0.456$, 0.00). So, reject H0. H1 is accepted.

H2 = There is a significant relationship between performance and satisfaction at the local Motorsport event. Performance and satisfaction of the spectator were moderately correlated, ($r = 0.328$, 0.00). So, reject H0. H2 is accepted.

H3 = There is a significant relationship between excitement and satisfaction at the local Motorsport event. Excitement and satisfaction of the spectator were moderately correlated, ($r = 0.460$, 0.00). So, reject H0. H3 is accepted.

H4 = There is a significant relationship between esteem and satisfaction at local Motorsport event Esteem and satisfaction of the spectator were moderately correlated, ($r = 0.294$, 0.00). So, reject H0. H4 is accepted.

H5 = There is a significant relationship between diversion and satisfaction at local Motorsport. Diversion and satisfaction of the spectator were moderately correlated, ($r = 0.370$, 0.00). So, reject H0. H5 is accepted.

4 Discussion

Spectators play a significant role in sport events. Having spectators' support would motivate the racer enthusiasm on games. Thus, the organizer needs to have more effort in order to maintain the loyalty of spectator to the local Motorsport event. SPEED Facet of motivation and satisfaction were important indicators to understand spectators. SPEED comprises of socialization, performance, excitement, esteem and diversion. This method was developed by past, researchers based of main motivation factor [5].

The finding of the relation between SPEED model and satisfaction in the setting of local motor sport can stand to gain this model allows motorsport marketers to fully comprehend what aspects should be new and unique in order to satisfy the spectators. Thus, the sport marketers can create a more meaningful marketing campaign in looking at socialization and excitement. The limitations of the present research provide opportunities for further research path to replicate and validate all or parts of the present research model, in order to determine the robustness of the findings in other sporting event settings [10].

Further research should also consider the importance of spectator loyalty towards motorsport as their most preferred sporting event. Perhaps, this prospective research endeavour could impact more interesting and deeper insights to both academicians and practitioners.

References

1. Correia, A., & Esteves, S. (2007). An exploratory study of spectators' motivation in football. *International Journal of Sport Management and Marketing, 2*(5–6), 572–590.
2. Cassidy, A. (2005). *A practical guide to information systems strategic planning*. CRC press.
3. Hirvonen, M. (2014). Motivational factors for sport spectator attendance: Case: Ice hockey.
4. Cordes, K. A., & Ibrahim, H. M. (1999). *Applications in recreation and leisure: for today and the future*. McGraw-Hill Book Company Europe.
5. Thien, H. T., & Van Mui, T. (2012). *Relationship between spectator's motivations and satisfaction in sporting events at Hochiminh City.* Paper presented at the 20th EASM Conference: Sport Between Business and Civil Society, Aalborg, Denmark: University College of Northern Denmark (UCN)(in co-operation with the Danish Institute for Sports Studies).
6. Kim, Y. K., & Trail, G. (2010). Constraints and motivators: A new model to explain sport consumer behavior. *Journal of Sport Management, 24*(2), 190–210.
7. Pease, D., & Zhang, J. (2001). Socio-motivational factors affecting spectator attendance at professional basketball games. *International Journal of Sport Management, 2*(1), 31–59.
8. Ross, S. D., Bang, H., & Lee, S. (2007). Assessing brand associations for intercollegiate ice hockey. *Sport Marketing Quarterly, 16*(2), 106.
9. Krejcie, R. V., & Morgan, D. W. (1970). Determining sample size for research activities. *Educ Psychol Meas*.
10. Van Leeuwen, L., Quick, S., & Daniel, K. (2002). The sport spectator satisfaction model: A conceptual framework for understanding the satisfaction of spectators. *Sport Management Review, 5*(2), 99–128.

Silat Tempur: The Combat Sports for Children

Mohamad Nizam Mohamed Shapie, Jamiaton Kusrin, Wahidah Tumijan and Mohd Shahiid Elias

Abstract The aim of the current study is to provide the review of Silat Tempur since it was introduced 2 years ago in Malaysia. Silat Tempur is a mini silat olahraga sports that focused to the children aged between 7 and 13 years old. The purpose of this combat sports is to improve the martial arts techniques for young exponents particularly to those who is still new in competition. The scoring points and rounds of a match is similiar as Silat Olahraga. The techniques used are punch, kick, catch, block and topple down which less than those reported by Shapie et al. (Journal of Combat Sports and Martial Arts 4:81–86, 2013 [1]) in Silat Olahraga. The high intensity and intermittent nature in Silat Olahraga (Shapie et al. in Performance analysis of sport VIII. Magdeburg, Germany, pp. 667–672, 2008 [2]) suggested that the competition is not appropriate to children. The arena of Silat Olahraga is big (9 × 9 m) compared to one straight line (4 × 9 m) for Silat Tempur. The arena of Silat Tempur is similar with fencing platform which aimed to ease the exponents to perform the simple silat techniques such as punching and kicking in competition. Moreover, the movements of both silat exponents can be controlled to four consecutive strikes for each exponent to avoid injury. This competition is in line with the national Seni Silat Malaysia curriculum (Shapie and Elias in Proceedings of the 1st World Congress on Health and Martial Arts in Interdisciplinary Approach, HMA 2015, p. 213, 2015 [3]) which emphasized on blocks, punches and kicks for the first two levels (lower phase; white to blue belt and blue to brown belt).

Keywords Seni silat malaysia · Silat olahraga · Combat sports · Gayung fatani

---

M. N. M. Shapie (✉) · J. Kusrin
Faculty of Sports Science and Recreation, Universiti Teknologi MARA (UiTM), Shah Alam, Selangor Darul Ehsan, Malaysia
e-mail: mohamedshapie@gmail.com

W. Tumijan
Faculty of Sport Science and Recreations, UiTM Seremban 3, Seremban, Negeri Sembilan, Malaysia

M. N. M. Shapie · M. S. Elias
Seni Gayung Fatani Malaysia Association, Shah Alam, Selangor Darul Ehsan, Malaysia

N. Sulaiman et al. (eds.), *Proceedings of the 3rd International Colloquium on Sports Science, Exercise, Engineering and Technology*,
https://doi.org/10.1007/978-981-10-6772-3_19

1 Introduction

Silat is a martial arts from the Malay ethnic group which is populated in mainland and islands Southeast Asia. Far from being merely a form of self-defense, martial arts are a pedagogical method that aims to create a certain culture and social ideals in the body of practitioner [4]. The word silat means a kind of sport or game, which consists of quick movements in attacking and defending [5]. Silat olahraga is a sport that existed in the midst of development of thousands of silat schools in Archipelago [5]. Olahraga means the ability for a silat exponent to perform his silat techniques in combat with striking and defensive actions such as punching, kicking, throwing, catching, parrying and blocking and any skill related to silat techniques [1].

The available literature on adult silat competition suggests that the sport is characterized by brief, high-intensity bouts of activity, short recovery periods and the need for competitors to repeatedly punch, kick and grapple with their opponent [1, 6, 7]. The overall intensity of a silat match indicated an exertion intensity close to the individual's maximal cardiovascular responses that were sustained throughout most of the match. This suggests that silat exponents are accustomed to numerous bouts of high force intensity actions alternating with lower intensity movements throughout a match [7]. Thus, it was suggested that the competition was not appropriate to children particularly to those who are still new in silat competition. It was reported that even in noncontact martial arts training, enthusiastic blocking technique may lead to bruising of the forearms in children [8]. The young silat exponents should be exposed step by step to silat competition based on their skills and level in silat [3].

The introduction of Silat Tempur competition since 2014 in Malaysia is to prepare the children in sparring format. It is a platform to establish and develop young athletes who have the courage, skilled in martial arts techniques and tactical in sports [9]. As young athletes (and their coaches) aspire to senior success it is important for them to have an appreciation of and be conditioned towards the demands of elite senior sport (silat olahraga) using Silat Tempur format. This paper will provide the review of Silat Tempur as a mini silat olahraga that focus to children.

2 The Sparring Competition in Silat

There are two sparring competition in silat; Silat Olahraga and Silat Tempur. The silat olahraga competition consists of three divisions according to age and weight either in male or female participants; youths (aged 12- to 14-year-old), teenager (aged 14- to 17-year-old) and adults (17-year-old and above) [10]. By the time silat exponents reach adolescence, it is likely they will be involved in combative competition. Here the adolescence may be in the period when many silat exponents make the transition from junior/teenager competition to compete in senior competition.

The Silat Tempur competition focus on children aged 7–13 years old. The objectives of this competition is to improve the techniques of punching, kicking, avoiding and blocking based on the Seni Silat Malaysia curriculum [3, 11]. The syllabus emphasized on blocks, punches and kicks for the first two levels of the curriculum (lower phase; white to blue belt and blue to brown belt). This competition provides early preparation for the children to master the basic techniques of silat before they switch to Silat Olahraga competition once they reach adolescence.

3 Silat Tempur

The purpose of this sport is to improve the silat techniques to children that still new to sparring format. The competition is based on platform that allowed the silat exponent to only move forward and backward during the match (Fig. 1). According to Anuar [12], movement is essentially a straight line either forward or reverse. Thus, the silat exponents need to master the basic attack and defense techniques in a straight line which can contribute to points during the match.

Silat Tempur used basic attack and defense methods on a straight line 2 × 9 m plantar. The platform equipped with flexible rubber mat (5 cm thick) to protect against injury. The competition derived from practical martial arts exponents that trained in pairs. Thus the movement of young silat exponents is controlled and limited to avoid injury during Silat Tempur match. The arena of Silat Tempur is smaller than Silat Olahraga (9 × 9 m) to ease the children to perform simple silat techniques such as kick and punch [5]. Silat Tempur has special rules that established to determine the game is safe in order to improve the quality of martial arts skills and inspiration [9] to young athletes. Silat instructor must provide training and understanding of the rules contending.

According to Shapie and Anuar [13] Silat Tempur has two categories for children and teenager to participate in this competition. The first category is **children aged from 7- to 13-year-old**. The male and female exponents start from class A (20–22 kg)

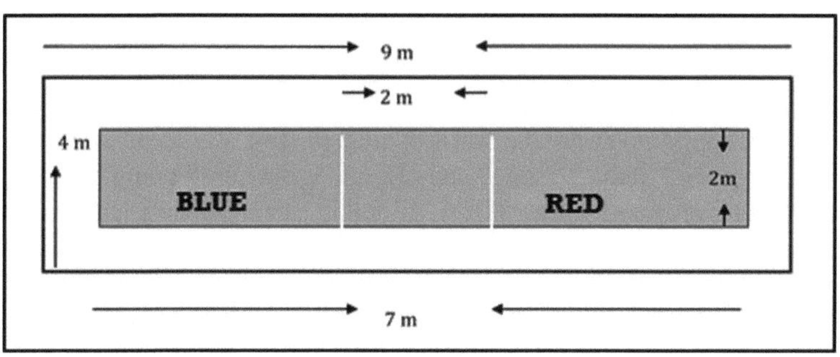

Fig. 1 The arena of Silat Tempur

Fig. 2 The Silat Tempur Competition

to free class (50 kg above). Overall, there are 11 classes for both male and females from age 7- to 13-year-old category, differentiated by 2 kg for each class. The second category is **teenager aged from 14- to 15-year-old**. The category starts from class A (26–30 kg) to free class (61 kg above). This category is differentiated by 5 kg for each class with eight classes for males and females.

In a Silat Tempur contest, two silat exponents square-off in a 7 m plantar area. The match consists of three rounds of 2 minutes each with 1-minute interval between rounds. This is a standard time to all age groups in silat competition. Time stoppages by the referee are not included in the actual bout time, so the actual duration of each round and thus total match time is usually longer than the 3 $\times$ 2 min scheduled, respectively.

Silat Tempur is simple silat competition [9] compared to silat olahraga [5, 14] which contains three main actions of kick, punch and topple down to score points. Points are awarded by the three judges for toppling an opponent, successful defensive blocks and offensive punches and kicks to the chest, abdomen and flanks, leg sweeps and throws [9]. Only strikes with either the arms or legs are considered legal. Unlike other competitive martial arts, the launch of any attack and defense movement must be initiated with specific coordinated silat "step patterns", otherwise points scored will not be valid. The silat step patterns are something that makes silat different to other martial arts. Every movement in a silat contest must be in silat posture (Fig. 2). It should not be similar like jumping in karate and taekwondo or the silat exponents will be panelised points. The exponent scoring the highest number of points or who knocks his opponent out wins [15].

The instruction is given by the wasit or referee (Fig. 2). There were two main instructions such as "Sedia, Pasang" and "Henti. Sedia" or "Pasang" define participants should be prepared while "Henti" is define as participants need to stop and prepared in each corner. The assumption is based on the rules of silat olahraga competition, which required the exponent to develop a competing pattern which consists of "sikap pasang" (silat posture), "pola langkah" (step pattern), measuring the distance against the opponent, coordination in performing an attack/defense, and finally return to "sikap pasang" [5]. "Rest" would typically be all the actions that happen in the round when the referee gives the command "henti" or "stop".

4 Techniques in Silat Tempur

Silat is a system that consists of positions and movements. When the athlete is moving in silat pola langkah, the positions and movements change continuously. As soon as one finds an opening target in their opponent's defense, he or she will try to finish the opponent with a fast attack. However in Silat Tempur competition there were five commans silat techniques used during the sparring such as punch, kick, block, catch and topple. All five silat exponent techniques can be defined as follows:

i. **Punch**: The punch "tumbuk" attack is done by a hand with a closed fist hitting the target. In silat, punching is often used to fight the opponent. It can be a straight punch "tumbuk lurus" or uppercut "sauk" to the exponent body's [5].
ii. **Kick**: The kick "tendang/terajang" is an attacking movement which is performed with one leg or two legs simultaneously. A kick can be aimed at any target. It can be front kick " tendang depan", sidekick "depak" or semi-circular sidekick "tendang lengkar" [5].
iii. **Block**: The blocking movements begin with the posture position "sikap pasang": the exponent stands straight with his hands around his body or close to his chest. Blocking or parrying "tangkisan" can be done using arms, elbows and legs with the purpose to block off or striking back at any attack [5].
iv. **Catch**: The catch "tangkapan" is done by using the hand to obstruct the opponent from carrying out an attack. The silat exponent is able to prevent himself from being attacked by pointing the attack which he has caught to another direction. A catch which twists or drags the opponent is forbidden. Also, a catch which could break the part which is being held such as the leg and waist is also forbidden. These regulations exist to protect the silat exponent's [5].
v. **Topple**: There are various ways of toppling down one's opponent. For example, a silat exponent "pesilat" can either push, shove the opponent's back leg from the bag or from the side, shove, hit, kick, strike or punch to make the opponent lose his balance. Every fall is considered valid as long as the silat exponent topples his opponent down without wrestling or he is able to overpower the opponent whom he has brought down [5].

Just like silat olahraga, there are several other skills that contribute to success in Silat Tempur performance where every aspect of these skills will contribute to success of the fighter. There are swipe, block, dodge/evade, topple down such as (i.e., scissors skills "*guntingan*"), jump, fake punch, fake kick and fake swipe. All of these skills will contribute on the attacking and defensive movements made by each fighter during the competition. However, not all of these techniques are necessary to the beginners in silat. As this is a subjective sport, it is important for each exponent to make a clear strike that can be observed by the judges. Thus, several factors will also influence the game such as injury and error of marking.

Silat competition rules stated that an exponent is allowed up to four consecutive punches and/or kicks to the opponent during a single attack, upon which the referee immediately breaks off the confrontation [2].

Table 1 Number of participations in Silat Tempur Competition 2014–2015

Tournament	Male Athletes (pax)	Female Athletes (pax)	Total (pax)
National Silat Tempur Competition 2014 (Universiti Malaya, Kuala Lumpur)	43	13	56
National Silat Tempur Competition 2014 (Universiti Teknologi MARA, Seremban 3, Negeri Sembilan)	63	27	100
National Silat Tempur Competition —Piala Aminuddin Anuar 2015 (Bandar Baru Bangi)	63	39	102
Total	169	79	248

5 Silat Tempur Competition

The first national Silat Tempur competition was held at Universiti Malaya, Kuala Lumpur Malaysia on April 11–13, 2014. The tournament was participated by 56 silat exponents from Kuala Lumpur, Selangor, Putrajaya, Melaka and Negeri Sembilan. The best male and female silat exponents in children category were Syamil Azim bin Nasahruddin (Kuala Lumpur) and Puteri Elissa Sarah Mohamad Nizam (Kuala Lumpur). While the best male and female silat exponents in teenager category were Mohamad Shazwan Arief (Selangor) and Nurul Nadia Fatin (Selangor).

The second national Silat Tempur competition was held at Universiti Malaya, Kuala Lumpur Malaysia on August 22–24, 2014. The tournament was participated by 100 silat exponents from Kuala Lumpur, Selangor, Putrajaya, Melaka, Johor and Negeri Sembilan. The number of participations increased compared to the previous competition due to the strong support from the Ministry (Table 1). The best male and female silat exponents in children category were Syamil Azim bin Nasahruddin (Kuala Lumpur) and Putri Nor Arisa (Negeri Sembilan). While the best male and female silat exponents in teenager category were Mohamad Shazwan Arief (Selangor) and Nurul Nadia Fatin (Selangor).

The third national Silat Tempur competition also known as Piala Aminuddin Anuar was held at Main Hall, Seksyen 7, Bandar Baru Bangi on 9–10 Mei 2015. The tournament was participated by 102 silat exponents from Kuala Lumpur, Perak, Selangor, Putrajaya, Melaka, Johor and Negeri Sembilan. The best silat exponents in children category was Puteri Elissa Sarah Mohamad Nizam (Kuala Lumpur). While, the silat exponents in teenager category was Mohamad Shazwan Arief (Selangor) and Nurul Nadia Fatin (Selangor). The overall champion for third consecutive times of national Silat Tempur competition was Selangor.

All three tournaments attracted many silat exponents either in the primary (7–12 years old) or secondary levels (13 years and above). The average fighter who competed was the new faces and senior who has competed since the first series of the game at the University of Malaya in April 2014.

6 Conclusion

Silat Tempur is a mini silat olahraga sport that specific to the children aged 7–13 years old. The purpose of this competition is to improve the silat techniques for young athletes that still new to competition. Moreover, it is a platform to prepare the young athletes before competing in silat olahraga. This is particularly true as the arena for Silat Tempur is smaller compare to silat olahraga. Thus, most athletes used five basic techniques for actions which were block, punch, kick, catch and topple down in spar. Not every technique that was reported by Shapie and co. [1] is necessarily implemented in the Silat Tempur competition.

Acknowledgements We would like to thank Akademi Silat Malaysia and Pertubuhan Seni Gayung Fatani Malaysia for volunteering for the study especially to Grandmaster Aminuddin Anuar his support in this study. Special thanks to Institute of Research Management and Innovation, Universiti Teknologi MARA for the financial support for this research (600-IRMI/DANA 5/3/LESTARI [0142/2016]).

References

1. Shapie M. N. M., Oliver, J., O'Donoghue, P., & Tong, R. (2013). Activity profile during action time in national Silat competition. *Journal of Combat Sports and Martial Arts, 4*(1), 81–86, January 2013. https://doi.org/10.5604/20815735.1073630.
2. Shapie, M. N. M., Oliver, J., O'Donoghue, P., & Tong, R. (2008). Distribution of fight time and break time in international Silat competition. In A. Hokelmann & M. Brummond (Eds.), *Performance analysis of sport VIII* (pp. 667–672). Germany: Magdeburg.
3. Shapie, M. N. M., & Elias, M. S. (2015). Seni Silat Malaysia: The Malay arts of self-defence. In R. M. Kalina (Ed.), Proceedings of the 1st World Congress on Health and Martial Arts in Interdisciplinary Approach, HMA 2015, Czestochowa, Poland (p. 213). Warsaw: Archives of Budo.
4. Wilson I. D. (2003). *The Politics of inner power: The practice of Pencak Silat in West Java*. Murdoch University.
5. Anuar, A. W. (1993). *Silat olahraga: The art, technique and regulations* (2nd ed.). Kuala Lumpur: Dewan Bahasa dan Pustaka.
6. Aziz, A., Tan, B., & Teh, K. C. (2002). Physiological responses during matches and profile of elite pencak silat exponents. *Journal of Sports Science and Medicine, 1*, 147–155.
7. Shapie, M. N. M., Oliver, J., O'Donoghue, P. G. & Tong, R. (2008, 3–6 September). *Distribution of fight time and break time in international silat competition*. Paper presented at the World Congress of Performance Analysis of Sport 8, Magdeburg, Germany.
8. Zetaruk, M. (2009). Children in combat sports. In R. Kordi, N. Maffulli, R. R. Wroble, & W. A. Wallace (Eds.), *Combat sports medicine*. New York: Springer.
9. Shapie, M. N. M., & Elias, M. S. (2014). The Silat tempur model: A comparison review with Silat Olahraga sport (Model Silat Tempur: Tinjauan Perbandingan Dengan Sukan Silat Olahraga). Kongres Warisan Melayu Sedunia 2014, MATRADE, Kuala Lumpur, Malaysia, October.
10. International Pencak Silat Federation. (2004). *Competition regulations*. Jakarta: International Pencak Silat Federation.

11. Anuar, A. W. (2002). Pendidikan dan Nilai-nilai Murni dalam Seni Silat (Education and Noble Values in Martial Arts). In Ministry of Culture, Art and Heritage, Manual jurulatih dan ringkasan mengajar (Instructor manual and summary of coaching), Tahap 1, Seni Silat Malaysia Untuk Jurulatih (First level, Seni Silat Malaysia for coaches). Paperwork. Kuala Lumpur: Ministry of Culture, Art and Heritage.
12. Anuar, A. W. (2007). Silat. Sejarah perkembangan kurikulum silat Melayu tradisi dan pembentukan kurikulum Silat Malaysia moden (Silat: The development history of traditional Malay silat and development of modern Silat Malaysia curriculum.). Bandar Baru Bangi, Selangor: Hizi Print Sdn Bhd.
13. Shapie, M. N. M., & Anuar, A. W. (2013). *Manual kursus pengenalan asas silat tempur*. Kuala Lumpur: Pertubuhan Seni Gayung Fatani Malaysia.
14. Shapie, M. N. M., & Elias, M. S. (2015). Silat Olahraga: The Malay combat sports. In: Kalina RM (Ed.), *Proceedings of the 1st World Congress on Health and Martial Arts in Interdisciplinary Approach, HMA 2015*, 17–19 September 2015 (p. 2012), Czestochowa, Poland. Warsaw: Archives of Budo.
15. Aziz, A. R., Tan, B., & Teh, K. C. (2002). Physiological responses during matches and profile of elite pencak silat exponents. *Journal of Sports Science and Medicine, 1,* 147–155.

Differences in Selected Performance Indicator Between Winning and Losing Team in Rugby Seven: Case Study on Vancouver World Rugby Seven Series 2015/2016 Season

Tuan Ainon Tuan Muda, Nurul Ain Muhammad Rafiai and Norasrudin Sulaiman

Abstract This cross-sectional study conducted to determine the differences in selected performance indicator between winning and losing teams in Vancouver World Rugby Series season 2015/2016. A total of 45 matches in this tournament were analyzed using hand notational analysis. The data analyzed using Statistical Package for Social Science version 19 (SPSS ver. 19) with the significance level was set at $p < 0.05$. Mann–Whitney Test was used to determine the differences in selected performance indicator between winning teams and losing teams. There were five significant indicators (try scoring, conversion, line out lost, line break, and turn over won) from the overall eighteen performance indicator that used in this study. Finding from this research will be suggested to coach to be implemented in training to prepare a winning team.

Keywords Performance analysis · Performance indicator · Rugby seven Winning team · Losing team · Team sport

1 Introduction

Seven a side rugby also known as rugby seven has developed since 1883 in Scotland, from recreation game and then become a competitive sport. It has been boosted by the Rugby World cup that was held in every 4 years and the annual 10 tournaments the International Rugby Board's World Rugby Seven Series that run from December

T. A. T. Muda · N. A. M. Rafiai · N. Sulaiman (✉)
Faculty of Sport Science and Recreation, Universiti Teknologi MARA, Shah Alam, Malaysia
e-mail: noras878@salam.uitm.edu.my

N. Sulaiman et al. (eds.), *Proceedings of the 3rd International Colloquium on Sports Science, Exercise, Engineering and Technology*,
https://doi.org/10.1007/978-981-10-6772-3_20

to June in the following years [1, 2]. Besides, in the upcoming Olympic Games Rio 2016, we will see the Olympic debut of Rugby Seven.

Rugby seven is a complex team sport that required combination of fitness and physical ability, highly execution of technical skill and tactical strategies for success in each tournament [3]. Recent studies also state that in rugby seven, the player is not much different between each other across the position, unlike the rugby union where there have differences between forward and backline player [4]. Match format also differs from rugby union, rugby seven was played by seven players per team and the tournament only played by 2 days with shorter playing duration 7 min per half with 2 min half time and for the final match, the game will be longer 10 min for each side in the final match. In contrast, rugby union played by 15 players on each side and the tournament usually will play up to seven days. The game will play for 40 min per half with 15 min rest time interval [3, 4].

In higher level of sporting event, the scientific and analytic support need to be investigate in order to improve and maximize the sporting performance [5] along with that identifying performance indicator in high-level sport are need to be perform stated by Hughes and Barlett [6] and James et al. [7]. Information about an individual or team performance can be a crucial factor either their team is winning or losing. Thus, sports performance analysis needs to be done in order to analyze either individual or team performance in order to increase and maximize the quality of rugby seven in all levels of the match. In order to increase the quality of the rugby seven, an identification which performance indicator that is influential in discriminate between winning and losing team needs to be done. Because of the performance analysis, the entire management team, coaches, and player can identify which strategies and tactical approach of their own team need to possess in order to succeed and also by analyzing opponent team strategy and tactical may give higher chances to win the game. An understanding of all indicators will contribute to the development of varied tactical approaches in this game and make rugby seven more exciting and complex in the future.

2 Research Design

The study will conduct to identify the differences in performance indicator between winning and losing team in Vancouver Rugby seven 2016. Total 45 matches were taken from official website of organizer [8]. A cross-sectional design will be used in this study and the video will analyze to get the result. Performance indicators that are chosen will be present as independent variables because the outcome will be based on individual and team performance itself. There were total 45 matches will analyze using hand national analysis system and will run the data using statistical analyze version 19.

3 Sampling

Subject for this study is all 16 teams that compete in HSBC Rugby World Seven Series: Vancouver, Canada. Team that participated was from Argentina, Australia, Brazil, Canada, England, Fiji, France, Kenya, New Zealand, Portugal, Russia, Samoa, Scotland, South Africa, United States of America and Wales. Video of a total of 45 matches from this tournament are analyzed from HSBC Rugby Seven official website [8]. A total of 45 matches were analyzed.

4 Instrumentation

Hand Notational Form

Hand notational form is used to mark the frequency of performance indicator happen during the video review. For this study, the researcher will use post-match video analysis where the video was taken from the official website. The video player and a laptop are needed to use during analysis of the video along with the hand notational form.

5 Data Collection Procedure

The total 45 matches ($n = 45$) were analyzed and the video was taken from the official website of rugby seven. A laptop with the video player and hand notational form are needed in order to analyze the video. The frequency of each indicator will be recorded in the form. They were various types of indicator related to the rugby. The indicators chosen in these studies were adopted based on the result of previous findings [1, 3, 9–11].

6 Results and Discussion

Table 1 below shows the selected performance indicators used in this study. While result of the mean and standard deviation between winning teams and losing teams in Vancouver World Rugby Series 2015/16 showed in Table 2.

Table 1 Selected performance indicator

No.	Performance indicator
1.	Tried score
2.	Conversion
3.	Scrum won
4.	Scrum lost
5.	Lineout won
6.	Lineout lost
7.	Penalty conceded
8.	Successful ruck
9.	Unsuccessful ruck
10.	Successful mauls
11.	Unsuccessful mauls
12.	Turnover won
13.	Passed completed
14.	Tackle made
15.	Tackle missed
16.	Knock on
17.	Forward pass
18.	Line break

7 Discussion

From the result, try score and conversion in winning team (try score 4.31 ± 1.743, conversion 2.71 ± 1.590) showed high significant difference compared to losing team (try score 1.56 ± 1.139, conversion 0.87 ± 0.842), in scrum, it show small difference in mean between winning team (scrum won 1.31 ± 1.104, scrum lost 0.13 ± 0.405) losing team (scrum won 1.33 ± 1.206, scrum lost 0.22 ± 0.517), for lineout also shows slightly difference in mean between winning team (line out won 1.09 ± 1.125, lineout lost 0.07 ± 0.252) compared to losing team (line out won 0.87 ± 0.842, line out lost 0.36 ± 0.712).

Next, in ruck and maul, there were no significant difference compared between winning team (successful ruck 7.24 ± 4.200, unsuccessful ruck 0.40 ± 0.720, successful maul 0.04 ± 0.208, unsuccessful maul 0.09 ± 0.288) and losing team (successful ruck 8.11 ± 4.041, unsuccessful ruck 0.76 ± 1.300, successful maul 0.09 ± 0.288, unsuccessful maul 0), in tackle made, penalty conceded, knock on and forward pass show least significant difference between winning team (tackle made 13.29 ± 5.802, penalty conceded 2.84 ± 1.609, knock on 0.64 ± 0.712, forward pass 0.27 ± 0.495) and losing team (tackle made 12.53 ± 6.148, penalty conceded 2.69 ± 1.807, knock on 0.80 ± 0.894, forward pass 0.29 ± 0.506).

Table 2 Descriptive data between winning and losing team

Indicator	Result	*N*	Mean	Std. deviation
Try score	Winning	45	4.31	1.743
	Losing		1.56	1.139
Conversion	Winning	45	2.71	1.590
	Losing		0.87	0.842
Scrum won	Winning	45	1.31	1.104
	Losing		1.33	1.206
Scrum lost	Winning	45	0.13	0.405
	Losing		0.22	0.517
Lineout won	Winning	45	1.09	1.125
	Losing		0.87	0.842
Lineout lost	Winning	45	0.07	0.252
	Losing		0.36	0.712
Successful ruck	Winning	45	7.24	4.200
	Losing		8.11	4.041
Unsuccessful ruck	Winning	45	0.40	0.720
	Losing		0.76	1.300
Successful maul	Winning	45	0.04	0.208
	Losing		0.09	0.288
Unsuccessful maul	Winning	45	0.02	0.149
	Losing		0	0
Tackle made	Winning	45	13.29	5.802
	Losing		12.53	6.148
Tackle missed	Winning	45	3.84	2.316
	Losing		5.36	2.715
Penalty conceded	Winning	45	2.84	1.609
	Losing		2.69	1.807
Passed completed	Winning	45	30.36	10.938
	Losing		26.16	11.872
Knock on	Winning	45	0.64	0.712
	Losing		0.80	0.894
Forward pass	Winning	45	0.27	0.495
	Losing		0.29	0.506
Line break	Winning	45	1.87	1.160
	Losing		0.56	0.693
Turnover won	Winning	45	1.04	1.127
	Losing		0.58	0.812

For tackle missed and passed completed it showed average significant in winning team (tackle missed 3.84 ± 2.316, passed completed 30.36 ± 10.938) compared to losing team (tackle missed 5.36 ± 2.715, passed completed 26.16 ± 11.872) and for a line break and turn over won data showed significant differences between winning (line break 1.87 ± 1.160, turnover won 1.04 ± 1.27) and losing team (line break 0.56 ± 0.693, turnover won 0.58 ± 0.812).

The winning team was excellent in try score and conversion where these two indicators are the main components to win the game. Winning team scores more on try and conversion compared to the losing team. It is indicated that utilizing the chance of giving conversion after a successful try is important, and it can be inferred that the winning team are stronger in offense play than the losing team. According to study by Ortega et al. [9] also found the same result where the try score and conversion were significantly different between winning and losing team and the study by Sulaiman et al. [10] showed only in try score was significant. While in lineout lost, it differs where the current result showed the positive result and previous study by Ortega et al. [9] showed there was no significance between the winning team and losing team.

Besides that, the winning team also found significantly higher in a line break and average in turnover won. Rugby seven has greater space, from the result shows winning team have more successful online break and it might be due to pattern play of each team, individual characteristic and the tactics used by the player on team. Turnover won demonstrate that the winning team is losing fewer balls in the play compare to losing team. In lineout lost, there is an average significance compared between both teams. It shows that winning team makes fewer mistakes in lineout lost because the losing team having greater mean compared to winning team. The study by Ortega et al. [9] also found significant differences in turnover won but the study does not included line break as their performance indicator.

In contrast, in set piece play both team shown no significant difference between winning and losing in the present study. Scrum won, scrum lost, lineout won, successful maul, unsuccessful maul, successful ruck and unsuccessful ruck associated with ball gaining phase in development of attacking play. This current result different stated by Ortega et al. [9]. Where the result showed the winning team significantly won more maul than losing team. The difference between the results may be related to the methodological, sampling variation, or the changing of pattern play in international rugby seven [12].

In tackling made, tackle missed, passes completed, penalty conceded, knock on, and forward pass shows no significant difference between both teams in ruby seven. Tackle is a defensive play in order to regain the ball possession in rugby seven shows no significant difference between winning and losing team and this might because they have greater space on one to another besides the fast phase of rugby seven unlike rugby union where the game are more compact [3]. Next, in penalty conceded also no significant and in passed completed, knock on and forward pass also does not have any significance in rugby seven Vancouver Rugby Seven Series.

Team performance indicator described in this study is formed of tactical and technical elements that have a relationship between chances of winning team. This study helps to clarify the element or indicators that differentiate the winning team.

Although this study outcome is different with previous study and it might differ with coaches belief, but the coaches must consider the multicomponent related to performance in rugby, the changes of the pattern play, and the also the level of competition itself.

8 Conclusion

As a conclusion, in this study, 45 matches Vancouver Rugby Seven Series were analyzed and the overall result showed certain performance indicators that is significant where try scoring, conversion, line out lost, line break, and turn over won are only significantly different between the winning and the losing team. It showed that in the higher competition, the team that fully seized the opportunity differentiates the team, whether their team becomes a winner or loser. The another 13 performance indicators used in this study were not significant scrum won, scrum lost, lineout won, successful ruck, unsuccessful ruck, successful maul, unsuccessful maul, tackles made, tackle missed, passed completed, knock on, forward pass, and penalty conceded.

The coaches may use this result of the study as references to monitor their team performance in the game. Besides, coaches and development team can develop the strategies and build more varying training system that will improve the team performance overall to increase the chances to win in each match. Athletes also can use this result to increase the individual performance by avoiding the behaviors or indicator that is higher in losing team. Official person also get benefits from this study where they can refer the current study to see the development and the quality, pattern play, where it will constantly change with time.

References

1. Hughes, M., & Jones, R. (2005). *Patterns of play of successful and unsuccessful teams in men's 7-a-side rugby union*, pp. 252–258.
2. Rooyen, M. K. V., Lombard, C., & Noakes, T. D. (2005). *Playing demands of sevens rugby during the 2005 rugby world cup sevens tournament*.
3. Higham, D. G., Hopkins, W. G., Pyne, D. B., & Anson, J. M. (2014). Performance indicators related to points scoring and winning in international rugby sevens. *Journal of Sports Science and Medicine, 13*, 358–364.
4. Ross, A., Gill, N., & Cronin, J. (2014). Match analysis and player characteristics in rugby sevens. *Sports Medicine (Auckland, N. Z.), 44*(3), 357–367. https://doi.org/10.1007/s40279-013-0123-0.
5. Vaz, L., Mouchet, A., Carreras, D., & Oria, H. M. (2011). The importance of rugby game-related statistics to discriminate winners and losers at the elite level competitions in close and balanced games. *International Journal of Performance Analysis in Sport, 11*, 130–141.
6. Hughes, M., & Barlett, R. M. (2002). The use of performance indicators in performance analysis. *Journal of Sports Sciences, 20*(10).

7. James, N., Mellalieu, S. D., & Jones, N. M. (2005). The development of position-specific performance indicators in professional rugby union. *Journal of Sports Sciences, 23*(1), 63–72. https://doi.org/10.1080/02640410410001730106.
8. World Rugby. (2016). *World Rugby HSBC Sevens Series*. http://www.worldrugby.org/sevens-series/video/events/1613#allmatches
9. Ortega, E., Villarejo, D., & Palao, J. M. (2009). Differences in game statistics between winning and losing rugby teams in the six nations tournament. *Journal of Sports Science and Medicine, 8*, 523–527.
10. Sulaiman, N., Azahan, A. A., Adnan, R., & Ismail, S. I. (2016). *Differences in game statistics between winning and losing teams in 2011 rugby world cup*, pp. 159–168. https://doi.org/10.1007/978-981-287-691-1_17
11. Vaz, L., Rooyen, M. V., & Sampaio, J. (2010). Rugby game-related statistics that discriminate between winning and losing teams in IRB and super twelve close games. *Journal of Sports Science and Medicine, 9*, 51–55.
12. Higham, D. G., Hopkins, W. G., Pyne, D. B., & Anson, J. M. (2014b). Relationships between rugby sevens performance indicators and international tournament outcomes. *Journal of Quantitative Analysis in Sports, 10*(1). https://doi.org/10.1515/jqas-2013-0095.

Marker-Less Motion Analysis of Turning Kick in Taekwondo

Shariman Ismadi Ismail, Nur Syazwani Abdu Razak and Norasrudin Sulaiman

Abstract This study presents a marker-less method to measure Taekwondo athletes' kicking velocity, and total kinetic energy generated from the movement. In this study, twelve ($n = 12$) national-level male Taekwondo athletes from three age categories (12–14 years old, 15–17 years old, and 18 years old and above) were recruited. All athletes were requested to perform turning kicks using back and front leg. Their movements were recorded using a marker-less motion analysis system that consists of a depth sensor (Kinect) and motion analysis software (Virtual Sensei Lite). The results reveal that the average kicking speed for the age categories of 18 years old and above meets the typical range of turning kick velocity that was reported in a previous study using conventional methods of measurement. In terms of the total kinetic energy produced during the movement, the average value is comparable to the value that was previously investigated using the conventional measuring methods. However, the usage of total kinetic energy data that were obtained in this study should be considered with caution. The large values of standard deviation in the data indicate that large variation existed in the data set. This is not the case with the kicking velocity data, which has shown significantly smaller standard deviation values.

Keywords Motion analysis · Marker less · Taekwondo · Kick

1 Introduction

Taekwondo is a sport originally from Korean martial arts. It is primarily used for self-defense, involving skillful techniques with powerful kicks directly to an opponent's head or upper trunk. This sport relies on speed, perfect agility, tactics, and power. The reaction time and response to an opponent are key elements to victory during a

S. I. Ismail (✉) · N. S. A. Razak · N. Sulaiman
Faculty of Sports Science and Recreation, Universiti Teknologi MARA, 40450 Shah Alam, Selangor, Malaysia
e-mail: shariman_ismadi@salam.uitm.edu.my

N. Sulaiman et al. (eds.), *Proceedings of the 3rd International Colloquium on Sports Science, Exercise, Engineering and Technology*,
https://doi.org/10.1007/978-981-10-6772-3_21

fight [1]. Taekwondo has been an official sport in the Summer Olympics since the year 2000 at Sydney, Australia.

According to Lee [2], the turning kick is the most common technique that is used to score points compared to other kicking, and is the fastest kicking technique in Taekwondo. The athlete performs a turning kick by the dominant or nondominant leg in a competition. The turning kick shows the highest peak kicking velocity among all kicks such as the chopping kick, back kick, and backswing [3].

It was previously investigated that the average maximum kicking velocity of the turning kick is between 12 and 16 m/s [4]. It was also previously reported that the impact force produced by the turning kick is around 1020 N [4]. However, there are no previous studies that investigate the kicking executed from the total amount of kinetic energy perspective. Total kinetic energy generated from a kicking motion can explain how much kinetic energy is generated from kick execution.

Measuring kinematics and kinetics parameters requires specific equipment that can be expensive. Conventional motion analysis requires marker-based infrared camera systems that are commonly equipped with a force plate system in order to perform kinematic and kinetic analysis. Recent advancements in motion analysis show that marker-less motion analysis is increasingly gaining the attention of researchers. This is because it is cost efficient [5–7] and reliable [8]. This study utilized the marker-less motion analysis method to investigate the turning kick movement in Taekwondo.

The main objective of this study is to demonstrate that a marker-less motion analysis system is capable of measuring important kinematics and kinetics parameters of the turning kick in Taekwondo.

2 Method

2.1 Participants

This study involves twelve ($n = 12$) national-level male Taekwondo athletes from three age categories (12–14 years old, 15–17 years old, and 18 years old and above). They were recruited from Majlis Sukan Negara (MSN) at Bukit Jalil among the junior and senior national teams, and also among the Bakat Team from Majlis Sukan Negeri Terengganu (MSNT). They are all physically healthy and free from any injuries. Consent was granted from the athletes and their parents (for athletes aged 17 and below) prior to the study. Four subjects were chosen for each age category.

2.2 Instrumentation

A marker-less motion capture system was used to record subjects' movements (Fig. 1). A Kinect depth sensor was used as the recorder. The depth sensor was integrated with a special software, Virtual Sensei Lite, in order to measure the kicking velocity and the total kinetic energy performed by the subjects. Based on the captured sequence of motion, the Virtual Sensei Lite produced the velocity and energy data. This data can then be transferred to other spreadsheet software for further analysis.

2.3 Procedures and Analysis

Participants were asked to perform three trials of turning kicks (Fig. 2) executed by the back leg, and another three trials by the front leg. The mean and standard deviation values were taken and split into three age groups.

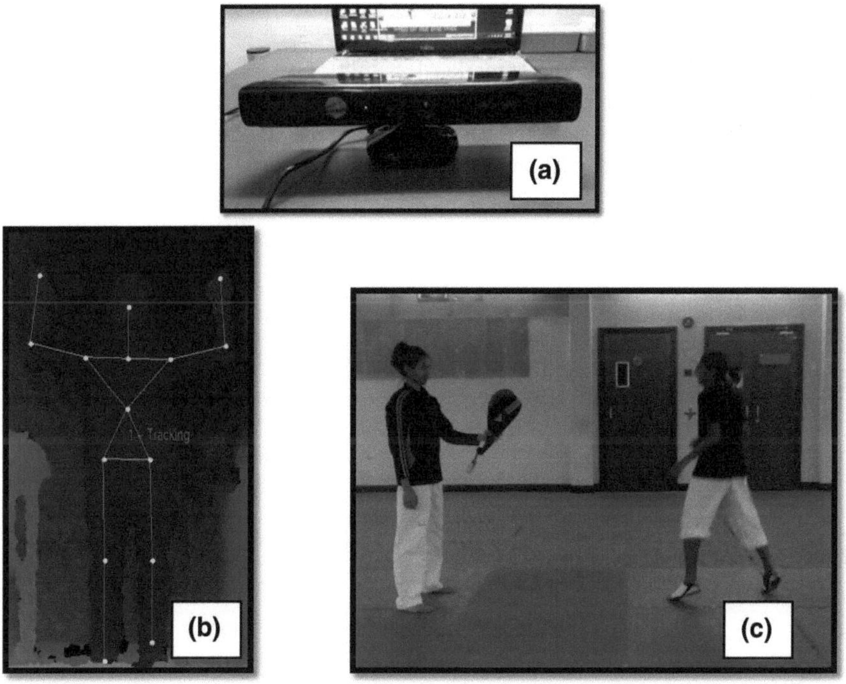

Fig. 1 **a** Depth camera. **b** Calibration procedure. **c** Kicking session

Fig. 2 Example of stages of turning kick execution

3 Results and Discussion

Below are the results of the study. Table 1 shows the average turning kick velocity (meter per second = m/s) for each age category, while Table 2 shows the average total kinetic energy (Joule = J) for each age category.

The results demonstrate that a marker-less motion analysis system is capable of measuring the kicking velocity and the total kinetic energy of the kicking movement. Based on Table 1, we can see the average turning kick velocity for age category 18 and above ranges between 12 and 14 m/s for both types of kicking. This range is

Table 1 Average turning kick velocity

Age category	Back kick vel. (m/s) (std. dev.)	Front kick vel. (m/s) (std. dev.)
12–14	9.06 (0.88)	9.24 (1.24)
15–17	9.37 (1.34)	9.86 (1.21)
18 and above	12.23 (2.19)	14.40 (3.10)

Table 2 Average total kinetic energy

Age category	Back kick TKE (J) (std. dev.)	Front kick TKE (J) (std. dev.)
12–14	1142.75 (203.4)	1639.75 (318)
15–17	1174.25 (50.43)	1821.50 (70.4)
18 and above	2315.25 (254)	2870.00 (274)

close to what was reported by Wasik [4], where the average kicking velocity of the turning kick was found to be between 12 and 16 m/s for adult Taekwondo athletes.

Table 1 also indicates that the average kicking velocity for age categories 12–14 and 15–17 are similar, in the range of 9–10 m/s for both kicking types.

This study also demonstrates that the measurement of the total kinetic energy produced during the kicking movement can be performed. The results show that the kinetic energy increases as the age increases. This is probably due to the increment of the athlete's mass as the age increases. The mass will influence the amount of total kinetic energy produced. However, the usage of the total kinetic energy data obtained from this study should be considered with caution. This is due to the large values of standard deviation across the data, indicating that a large variation exists in the data set. This is not the case with the kicking velocity data, which shows significantly smaller standard deviation values.

Meanwhile, it is difficult to compare the results between the total kinetic energy obtained in this study with other studies because of different measurement values. In order to create a feasible comparison, several assumptions need to be established. Previous studies [4] highlight that the impact force produced during the turning kick is around 1020 N. If assumptions were made that the average male height is 173.1 cm [9], and the turning kick effective length is based on the leg length [10], which represents 24.7% of the total body height, we can estimate that the effective distance of the kicking leg during the turning kick is around 50 cm. If we use this value and assume that the impact force during the turning kick is around 1020 N, it can be estimated that the impact force of 1020 N was produced with an average energy of 510 Nm (510 Nm = 510 J) during the impact phase. The total kinetic energy obtained from the turning kick of age group 18 and above is in the range of 2300–2800 J. The larger amount of average total kinetic energy during the kicking execution when compared to the energy that was produced during the impact phase is logical because there are energy losses during the impact phase. There is a need to estimate the kicking efficiency, similar to the concept of the coefficient of restitution during turning kick impact, should this scenario need further clarification.

4 Conclusion

This study provides a low-cost method to measure kicking velocity and total kinetic energy generated from a Taekwondo turning kick motion. The kicking velocity data indicates that this method is acceptable. However, the total kinetic energy data suggests that further clarification is still required in order to confirm its reliability and validity.

Acknowledgements This work was sponsored by Universiti Teknologi MARA [600-RMI/DANA 5/3/ARAS (54/2015)]. We also would like to thank the Faculty of Sports Science and Recreation, Universiti Teknologi MARA, Malaysia for supporting this study.

References

1. Asia, A. A., & Warkar, A. B. (2013). Auditory and visual reaction time in taekwondo players. *International Journal of Recent Trends in Science and Technology, 8*(3), 176–177.
2. Lee, J. B. (1998). *A study of kicking techniques of advanced Korea taekwondo players* [Unpublished Coaching Report, Korea Institute of Sport Science, Seoul].
3. Young, K. K., Yoon, H. K., & Shin, J. (2011). Inter-joint coordination in producing kicking velocity of taekwondo kicks. *Journal of Sports Science and Medicine, 10,* 31–38.
4. Wasik, J. (2011). Kinematics and kinetics of taekwon-do side-kick. *Journal of Human Kinetics, Section I-Kinesiology, 30,* 13–20.
5. Timmi, A. (2011). Virtual Sensei Lite. http://www.virtualsensei.it/lite/. Last accessed: August 16, 2016.
6. Bünger, M. (2013). *Evaluation of skeleton trackers and gesture recognition for interaction.* Denmark: Aalborg University.
7. Mustapha, G., Zakaria, M., Sulaiman, W. R. W., & Mahmud, J. (2014). The mechanical aspects of martial arts : Total time of execution and kinematics of Kaedah A. In: R. Adnan, S. I. Ismail, & N. Sulaiman (Eds.), *Proceedings of the International Colloquium on Sports Science, Exercise, Engineering and Technology 2014 (ICoSSEET 2014)* (pp 3–12). Singapore: Springer.
8. Tuan Yusuf, K. M. S., Ahmad Nazri, A. F., Mustapha, G., & Mahmud, J. (2015). Analysis of static and dynamic motion accuracy for kinect-virtual Sensei system. *ARPN Journal of Engineering and Applied Sciences, 10*(17), 7328–7335.
9. Leva, P. (1996). Adjustments to Zatsiorsky-Seluyanov's segment inertia parameters. *Journal of Biomechanics, 29*(9), 1223–1230.
10. Plagenhoef, S., Evans, F. G., & Abdelnour, T. (1983). Anatomical data for analysing human motion. *Research Quarterly for Exercise and Sport, 54,* 169–178.

Differences in Variable Goal Scoring Characteristics Between Winning and Losing Team in AFF Futsal Tournament 2014

Norasrudin Sulaiman, Siti Hartini Azmi and Shariman Ismadi Ismail

Abstract This study aims to analyze and investigate the differences in goal scoring characteristics among winning and losing teams in the AFF Futsal Tournament, 2014. The independent variable (IV) represents the two separated groups, winning and losing teams; while the dependent variable (DV) represents the variable goal scoring characteristic. The sample is divided into two groups that consist of 24 samples (matches). In this study, the Wilcoxon signed-rank test or nonparametric test was used to make a comparison between the two sets of scores that come from the same participants, which are winning teams and losing teams, which come from the same group. The purposive sampling technique was used to obtain a sample from the target population of the study. The subjects were all selected from the AFF Futsal Tournaments, 2014. The analysis results show that there is a significant difference in terms of goal scoring characteristics based on open play ($p < 0.05$, for seven out of eight variables) and set play ($p < 0.05$, for two out of five variables) situation when compared between the winning and the losing team. Due to these results, teams in the winning group have a better gameplay, good game technicals, and tacticals. Meanwhile, the teams in the losing group do not have good tacticals and need to give more attention to their weaknesses.

Keywords Performance analysis · Performance indicators · Futsal · Goal scoring characteristics

N. Sulaiman (✉) · S. I. Ismail
Faculty of Sports Science and Recreation, Universiti Teknologi MARA, 40450 Shah Alam, Selangor, Malaysia
e-mail: noras878@salam.uitm.edu.my

S. H. Azmi
Faculty of Sport Science and Coaching, Universiti Pendidikan Sultan Idris, Tanjung Malim, Perak, Malaysia

N. Sulaiman et al. (eds.), *Proceedings of the 3rd International Colloquium on Sports Science, Exercise, Engineering and Technology*,
https://doi.org/10.1007/978-981-10-6772-3_22

1 Introduction

Futsal is a sport that involves five-versus-five players in an indoor soccer match. This sport was introduced in 1930 in Uruguay and Brazil. The aim of this sport is slightly different from football, as it is played in restricted spaces [1]. This sport is also organized by FIFA, similar to football. This sport is played by both men and women. There are some specific characteristics that are required for those involved in this sport because Futsal is an intermittent sport that is demanding of high physical aspects, good technical ability, and also has tactical demands on players.

Performance analysis is an important way to use or analyze sports for giving feedback on whether there are positive or negative effects of performance by the team, or individual skills. The performance analysis can be carried out in several ways to determine weaknesses, or to improve the overall performance of the athletes. Some sports use feedback through videos and computerized versions of performance analysis. Performance analysis can also cover almost all the factors related to the game, so the coach and the athletes can observe their positive and negative actions through the game, and use this to improve performance. Some of the analyses that can be analyzed by the performance analysis are strategies, playing patterns, scoring points, etc. [2].

Nowadays, sport organizations and coaches use performance analyses to enhance the performance of their players and teams. The performance analyses are usually used by the coach to know the weaknesses and strengths through related equipment such as video cameras and computerized software. Performance analyses can also be used by the coach as a platform to enhance matches and strategies. The combinations of sports and technologies nowadays are becoming popular or highly demanding, and are mostly used by professional teams in all sports such as football, hockey, and basketball.

Futsal has become a fast sport because the movement of the game is highly demanding of physical fitness. Besides that, the players are free to change to substitute and play again. So, the players usually have some time to recover and the team will remain in the fast phase of the game. Performance analyses can be a medium to the coach or athletes to enhance their performance. The coach will take the results from the performance analysis such as game plays, strategies, weaknesses, advantages of their team, and can then improve through their knowledge of the game so the players or team can change a pattern or performance to obtain better results.

2 Method

2.1 Research Design

In this study, the type of research design was a descriptive analysis study. In this study, the researcher used the Wilcoxon Signed-Rank Test using SPSS Statistics to

Table 1 Selected performance indicators

Groups of variables	Variables
Variables related to winning teams	Set play, open play, whom, from a distance
Variables related to losing teams	Set play, open play, whom, from a distance

gather, identify and measure the data. This test was conducted at the field and at a laboratory. The field involved is Stadium Malawati Shah Alam. The laboratory test was conducted at the Performance Analysis Laboratory of Faculty of Sport Science and Recreation of UiTM Shah Alam. The laboratory was chosen because it has a complete infrastructure for sport coding. A number of 10 teams from Asian country Futsal teams were involved, with a total of 24 matches. The researcher recorded the data after every game.

2.2 *Sample Selection*

The purposive sampling technique was used to obtain the samples from the target population of the study. The subjects are participants in all the AFF Futsal Tournaments, 2014. The subject and teams have the criteria that the researcher needs; they are all professional Futsal players. The participants were selected from 10 teams that joined the AFF Futsal Tournaments, 2014. The results from the tournaments were separated into two groups, winning and losing teams. The researcher found that this technique was the most appropriate. A number of 24 matches were used for the data; all of the matches covered all the games of the tournaments.

2.3 *Data Analysis*

All the matches were recorded and transferred into the Sport Code plus (Sportstec limited, Australia) software for analysis. All the four indicators were divided into two groups of variables. Next, all the data gathered were transferred into the Statistical Package for the Social Sciences (SPSS) version 19.0 for analysis. All the performance indicators are shown in Table 1.

Table 2 Goal scoring related to open play between winning and losing teams

Groups of variables		Winners		Losers		
		Mean	SD	Mean	SD	*p*
Open play	Ala play	0.54	0.884	0.08	2.82	0.032
	Combination	2.17	1.761	0.21	0.658	0.000
	Df splitting distribution	1.96	1.967	0.42	0.881	0.001
	GK distribution	0.17	0.381	0.00	0.00	0.046
	Solo effort	1.13	1.116	0.13	0.338	0.001
	Rebound	0.87	0.947	0.17	0.381	0.002
	Power play	0.21	0.658	0.00	0.00	0.102
	Counterattack	0.54	0.884	0.00	0.00	0.010

3 Results

3.1 *Open Play*

Illustrative information for variables of goal scoring related to open play between winning and losing teams were gathered and investigated for every game. The mean and standard deviation values are introduced in Table 2.

From data gathered in the table, a summary of the results is presented. A total of 24 ($n = 24$) subjects of matches were used in this research. 24 represents both winning and losing teams. From the data collected, it shows that the measurement of mean (M) and standard deviation (SD) for variables related to the open play of goal scoring are related to both winning and losing teams. The variables in open play are included ala play, combination play, defender splitting distribution, goalkeeper, distribution, solo effort, rebound, power play, and counterattack. The mean and SD for ala play are (±0.54, ±0.884 and ±0.08, ±2.82). The mean and SD for combination play are (±2.17, ±1.761 and ±0.21, ±0.658). The mean and SD for defender splitting distribution for both groups are (±1.96, ±1.967 and ±0.42, ±0.881). The mean and SD for goalkeeper distribution for winning and losing teams are (±0.17, ±0.381 and ±0.17, ±0.00). The mean and SD for solo effort for winning and losing teams are (±1.13, ±1.116 and ±0.13, ±0.338). The mean and SD of rebound for winning and losing are (±0.87, ±0.947 and ±0.17, ±0.381). The mean and SD of power play for winning and losing are (±0.21, ±0.658 and ±0.00, ±0.00). For the last variable for open play, which is counterattack, the mean and SD for winning and losing are (±0.54, ±0.884 and ±0.00, ±0.00).

Table 3 Goal scoring related to set play between winning and losing teams

Groups of variables		Winners		Losers		*p*
Set play		Mean	SD	Mean	SD	Mean
	After corner kick	0.29	0.464	0.04	0.204	0.034
	Direct free kick	0.13	0.338	0.00	0.00	0.083
	Following free kick	0.21	0.509	0.08	0.282	0.564
	Kick in	0.04	0.204	0.04	0.204	1.000
	Penalty	0.17	0.381	0.000	0.000	0.046

3.2 *Set Play*

Illustrative information for variables of goal scoring related to set play between winning and losing teams were gathered and investigated for every game. The mean and standard deviation values are shown in Table 3.

From the data that was gathered from Table 3, a summary is presented. The total subjects in this study is 24 matches ($n = 24$), which are divided into five winning and five losing set play indicators.

The data collected shows the measurements of winning and losing of mean (M) and standard deviation (SD) for variables in set play of after corner kick, direct free kick, following free kick, kick in, and penalty. The mean and losing for winning and losing of set play after corner kick are (±0.29, ±4.64 and ±0.4, ±0.204); and for direct free kick is (±0.13, ±0.338 and ±0.00, ±0.00). The mean and SD for set play in following free kick are (±0.21, ±0.509 and ±0.08, ±0.282). The mean and SD for kick are (±0.04, ±0.204 and ±0.04, ±0.204). Meanwhile, for set play in penalty for winning and losing are (±0.17, ±0.381 and ±0.000, ±0.000).

4 Discussion

The first objective is to investigate the variables related to the goal scoring characteristics between winning and losing teams in AFF Futsal Tournament, 2014, based on open play. Mentioned that the performance analysis needs to be carried out in sports to help improve performance. Performance analysis is becoming an important aspect nowadays for sports, especially when relates to team sports. With the existence of this performance analysis, competition of Futsal has become more interesting and competitive. Performance analysis also can help with the development of this sport. Open play consists of game tacticals, which the players or teams create a goal with the condition of the ball in a continuous move.

Based on the data collected, there is a significant difference of variables related to goal scoring based on open play by winning and losing teams.

The second objective of this study is to investigate the variables related to goal scoring between winning and losing teams based on set play. Goals from set play can be defined as goals that are scored with the condition of the dead ball situation. Examples of the set play in Futsal are penalty, direct free kick, following free kick and also corner kick. Based on the data collected, the winning teams group achieved a greater score compared to the losing teams. It can be related or effected by the tacticals of the team. A good team will have a better of pattern gameplay. This can be due to proper training with lot of variation of movement. With match analysis, teams can study how to optimize the training process so their players can improve their weaknesses and increase their strength [3].

Previous studies revealed that Futsal is a sport that is physically demanding [4]. In this study, distances of the goals measured included goals that come from inside the penalty area, between the penalty mark and goal line, between penalty mark and second penalty mark, between second penalty mark and halfway line, and from own half. This investigation is important to both winning and losing teams. It is because this can help the coach know the most successful distances used to score a goal. For the losing team, it also can help the team to know their weaknesses. For example, the side or distance that usually made mistakes and gave the goal to the opponents can be known and given more attention. Based on the data collected, there is a significant difference between winning and losing teams in terms of goal scoring based on the distance of goals scored.

The last objective of this study is to investigate the difference of scorers, which is between field player and field players. Field player consists of the whole team in the court, except for goalkeeper. From the data collected, both teams have a no significant difference on goal scorer. Both teams, as usual, have the highest number that scores the goal from their field players. This investigation shows that goalkeeper does not have the ability or chance to be a scorer.

5 Conclusion

In conclusion, this study examined the differences of goal scoring characteristics of winning and losing Futsal teams in the AFF Futsal Tournament, 2014. This investigation allowed the researcher to investigate the statistical data of winning and losing, which can deliver beneficial info for Futsal development. The analysis results show that there is a significant difference in terms of goal scoring characteristics based on open play, set play, and distance of scoring a goal. Due to these results, teams in the winning group have a better gameplay, good game technicals, and tacticals.

Meanwhile, the teams in the losing group do not have good tacticals and needs to give more attention to their weaknesses. A use of other aspects can also help the team, for example, performance analysis. This medium can help the team to know their opponent's strengths and their own weaknesses, so the players and also the coach can improve their overall performance.

Acknowledgements This work was supported by the Malaysian Ministry of Education's Sports Grant (Geran Sukan KPT) [KPT.N.660-7 Jld 8 (57) and 100-RMI/GOV 16/6/2 (3/2014)]. We also would like to thank the Faculty of Sports Science and Recreation, Universiti Teknologi MARA, Malaysia for supporting this study.

References

1. Castagna, C., Chaouachi, A., Rampinini, E., Chamari, K., & Impellizzeri, F. (2009). Aerobic and explosive power performance of elite Italian regional-level basketball players. *The Journal of Strength and Conditioning Research, 23*(7), 1982–1987.
2. Norasrudin, N. S., Redzuan, S., & Mubin, A. P. (2012). Distinguishing playing pattern between winning and losing field hockey team in Delhi FIH road to London 2012 tournament. *International Scholarly and Scientific Research & Innovation, 6*.
3. Hughes, M. D., & Franks, I. M. (2004). *Notational analysis of sport 2nd edition-better systems for improving coaching and performance*. London: E. & F.N. Spon.
4. Leite, W. (2016). Physiological demands in football, futsal and beach soccer: A brief review. *European Journal of Physical Education and Sport Science, 2,* 1–10. https://doi.org/10.5281/zenodo.205160.

Reflection Rate Index of Passive Markers for Motion Capture Application Based on Different Colors and Sizes

Shariman Ismadi Ismail, Muhammad Fazrul Faiz Samsudin and Norasrudin Sulaiman

Abstract Motion capture and motion analysis are employed in animation development and modeling of body movements in various fields. A typical motion capture system requires markers to identify the movement performed by objects in motion that is being tracked. A passive marker is built from light-reflecting materials, and the motion capture systems require lighting during the motion capture. The purpose of this study is to examine the influence of colors and sizes of markers on its reflection rate index (RRI). This study involved the RRI of 12 different markers with various sizes (10, 14, and 18 mm) and colors (blue, red, gold, and silver). Based on the RRI value, the type and level of reflection and light distribution from each marker were identified. The results reveal that silver marker has an RRI value above 1, while blue, red, and gold markers have an RRI value below 1. Based on this study, we can categorize each marker's light reflection rate based on the calculated RRI. This is helpful to users in deciding what type of marker needs to be used in each respective area.

Keywords Motion capture · Marker · Reflection · Color · Size

1 Introduction

We have witnessed the application of motion capture in various areas since the nineteenth century [1]. The increasing demand of motion capture applications caused these technologies to expand their systems specifications to meet specific requirements for each study that utilized the systems.

The validity and reliability of a marker-based motion capture have been reviewed extensively in previous works [2]. If the systems are properly utilized, we could obtain reliable data.

S. I. Ismail (✉) · M. F. F. Samsudin · N. Sulaiman
Faculty of Sports Science and Recreation, Universiti Teknologi MARA, 40450 Shah Alam, Selangor, Malaysia
e-mail: shariman_ismadi@salam.uitm.edu.my

N. Sulaiman et al. (eds.), *Proceedings of the 3rd International Colloquium on Sports Science, Exercise, Engineering and Technology*,
https://doi.org/10.1007/978-981-10-6772-3_23

The art motion capture systems require markers to identify movements performed by objects in motion that are being tracked. These markers come in many different sizes according to the requirements of the industries. The markers that are used in motion capture systems could be either active markers, where they have lighting capabilities with either conventional lamps or light-emitting diodes (LED); or passive markers, where the markers are built from light-reflecting materials and the motion capture systems require lighting during the motion capture. Both active and passive marker systems are suitable for either two-dimensional or three-dimensional motion capture setups, as long as the motion analysis system can interpret the coordinates of the tracked objects that have been identified by the markers used in the motion capture system [3].

In terms of the passive marker, there was evidence that, under certain circumstances, the identification process of the marker location is difficult. One of the situations includes the scenario, whereby two markers are located very close to each other [4]. It was suggested that this is due to the redundancy of the reflection from both markers [3].

Previous studies attempted to measure and categorize the reflection rate of markers for motion capture by identifying its Reflection Rate Index [3]. Some studies indicated that the marker surface condition could potentially influence the reflection behavior [5, 6]. However, no study has been conducted in order to understand the influence of the marker color and size on the Reflection Rate Index. Motivated by this research gap, the purpose of this study is to examine the influence of colors and sizes of markers on its Reflection Rate Index (RRI). The RRI of 12 markers with various sizes (10 mm, 14 mm, and 18 mm) and colors (blue, red, gold, and silver) were measured in this study.

2 Method

2.1 Sample

In this study, 12 reflective markers with various sizes (diameter of 10, 14, and 18 mm) and colors (red, gold, blue, and silver) were utilized. This is shown in Fig. 1.

2.2 Instrumentation

Data collection was performed in a dark room. The markers are located parallel with respect to the center of the camera lens (Olympus TG-3). The distance between the lens and the marker surface is 5 cm. A light source was utilized in order to generate a constant reflection from the marker (Fig. 2).

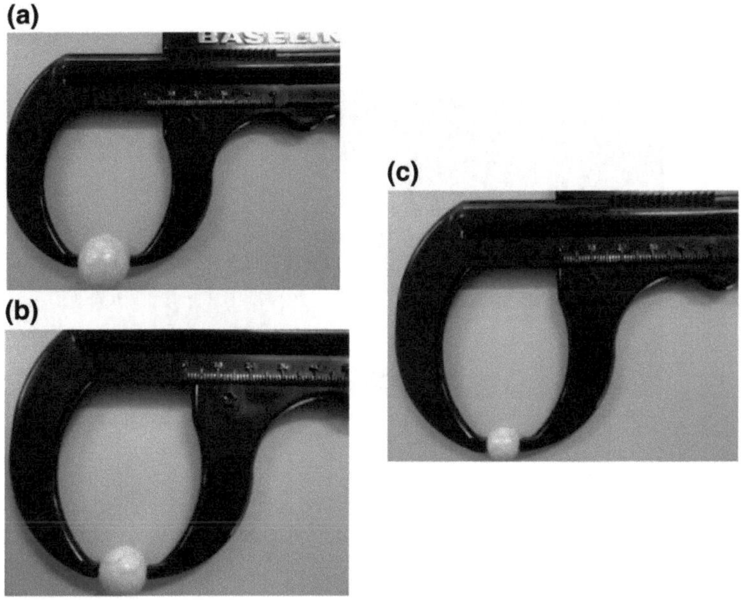

Fig. 1 Marker diameter: **a** 18 mm, **b** 14 mm, **c** 10 mm

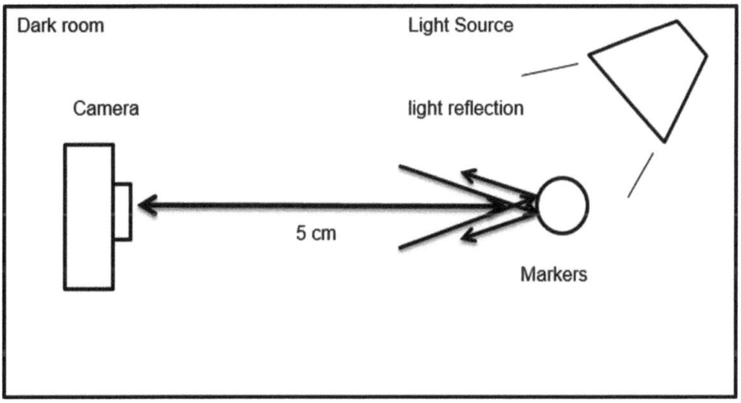

Fig. 2 Instrumentation setup

2.3 Analysis

Each marker reflection condition was recorded by the camera to obtain its still image. Each of the 12 markers' still images was then processed by using Kinovea (v0.8) software to measure the Reflection Rate Index (RRI) (Fig. 3) [3]. Each marker's RRI was calculated 30 times, and the average value was obtained. The RRI was measured based on the following relationship [3]:

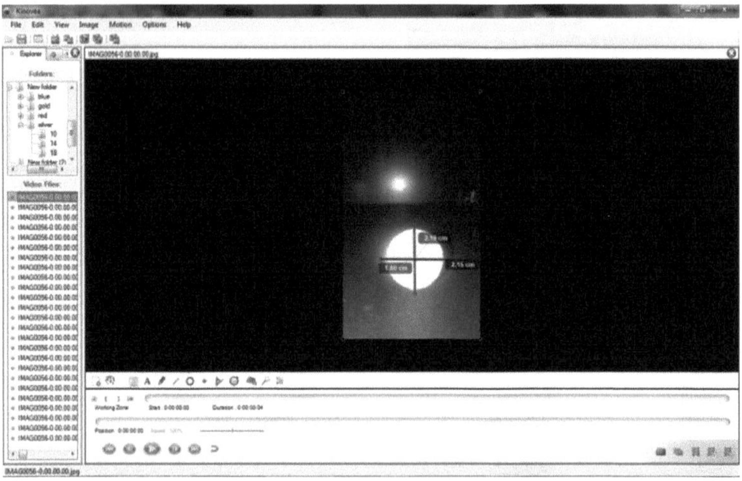

Fig. 3 RRI measurement with Kinovea

Table 1 RRI average values for all markers

Colors	Sizes		
	10 mm	14 mm	18 mm
Blue	0.87	0.89	0.85
Red	0.81	0.84	0.79
Gold	0.77	0.76	0.73
Silver	1.68	1.64	1.55

$$\mathrm{RRI} = \frac{\text{total glowing light cross section area}}{\text{total cross section area of marker}} \tag{1}$$

Comparison of the RRI for each marker was then conducted in order to observe the influence of marker size and color on its RRI.

3 Results and Discussion

The results obtained in this study are shown in Table 1. The result demonstrates that the blue markers have RRI value smaller than 1 for the three sizes; while the silver markers have the largest values compared to the other markers. This can also be observed from the reflection condition shown in Fig. 4.

There was an increase in the RRI value for both blue and red markers from 10 mm to 14 mm in size (0.87–0.89 and 0.81–0.84, respectively). However, there was a decrease in RRI value from 0.89 to 0.85 and 0.84 to 0.79 for the blue 18 mm marker,

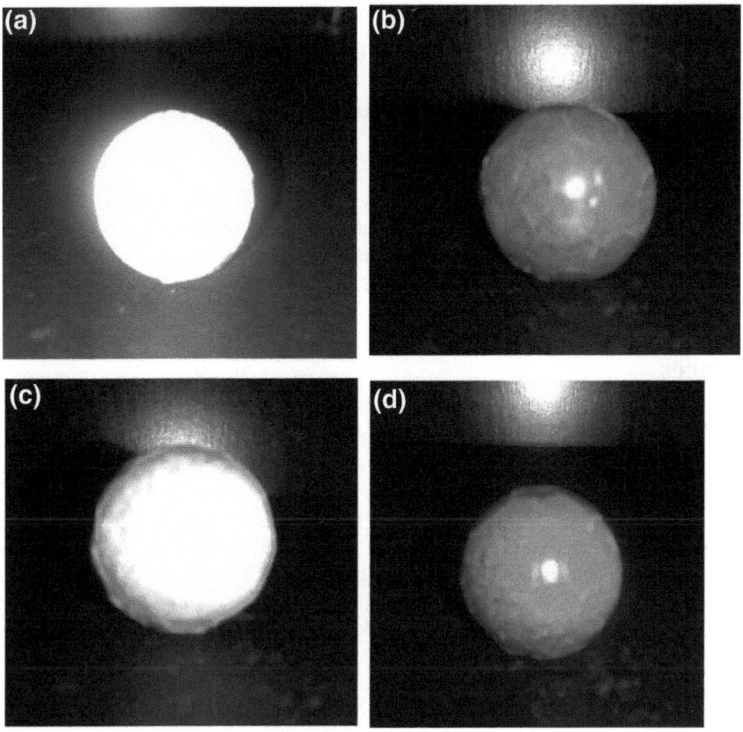

Fig. 4 Example of Reflection condition: **a** silver, **b** red, **c** gold, **d** blue

and red 18 mm marker, respectively. Gold and silver markers on the other hand, show a decrease in RRI value continuously from 0.77–0.76 to 0.73, and 1.68–1.64–1.55, respectively, based on 10, 14, and 18 mm sizes. It was noted that there is no significant difference in the reflection rate index with regards to different sizes of the markers.

4 Conclusion

Based on the results, the silver color shows the highest RRI, which remains over 1, for all three sizes. The natural frequency of the silver color with a wavelength of 412 nm could potentially contribute to this effect. The silver color has good reflectivity that does not vary with wavelength, and thus appears very close to white. The absorption of the peak of the silver color lies in the ultraviolet region. Consequently, silver maintains high reflectivity evenly across the visible spectrum, and looks pure white. The sizes of the markers do not show any effect on the RRI value of the markers.

This study reveals that different marker colors could potentially provide different reflection properties. Thus, users can determine the type of reflection that is suitable for various scenarios. There are users who prefer lower RRI values, while others prefer higher RRI values, according to their motion capture settings. There are situations where higher values of RRI are not suitable. Thus, users will choose color markers with an RRI range of optimal performance, based on the settings and scenarios available.

Acknowledgements This work was sponsored by Universiti Teknologi MARA [600-RMI/DANA 5/3/ARAS (54/2015)]. We also would like to thank the Faculty of Sports Science and Recreation, Universiti Teknologi MARA, Malaysia for supporting this study.

References

1. Sharma, A., Agarwal, M., Sharma, A., & Dhuria, P. (2013). Motion capture process, techniques and applications. *International Journal on Recent and Innovation Trend in Computing and Communication, 1*(4), 251–257.
2. Lempereur, M., Brochard, S., Leboeuf, F., & Remy-Neris, O. (2014). Validity and reliability of 3D marker based scapular motion analysis: A systematic review. *Journal of Biomechanics, 47*(10), 2219–2230. https://doi.org/10.1016/j.jbiomech.2014.04.028.
3. Ismail, S. I., Adnan, R., & Sulaiman, N. (2014). Reflection rate index of markers for motion capture application. In: R. Adnan, S. I. Ismail, & N. Sulaiman (Eds.), *Proceedings of the International Colloquium on Sports Science, Exercise, Engineering and Technology 2014 (ICoSSEET 2014)* (pp. 21–28). Singapore: Springer.
4. Richards, J. G. (1999). The measurement of human motion: A comparison of commercially available system. *Human Movement Science, 18,* 589–602.
5. Miranda-Medina, M. L., Somkuti, P., & Steiger, B. (2013). Detection and classification of orange peel on polished steel surfaces by interferometric microscopy. *Journal of Physics: Conference Series, 450,* 1–6. https://doi.org/10.1088/1742-6596/450/1/012009.
6. Kigle-Boeckler, G. (1996). Measuring gloss and reflection properties of surfaces. *Tappi Journal*, 194–198.

MIX
Papier aus verantwortungsvollen Quellen
Paper from responsible sources
FSC® C105338

If you have any concerns about our products,
you can contact us on
ProductSafety@springernature.com

In case Publisher is established outside the EU,
the EU authorized representative is:
Springer Nature Customer Service Center GmbH
Europaplatz 3, 69115 Heidelberg, Germany

Printed by Libri Plureos GmbH
in Hamburg, Germany